CULTURAL TECHNOLOGIES

Cultural Technologies: Robots and Artificial Intelligence in the Performing Arts presents a diverse range of perspectives from leading scholars and artists on contemporary performing arts practices that engage with robotic and AI (artificial intelligence) technologies.

In Part One, "Robot/AI Cultures and Performing Arts Practices," contributors discuss how cultural understandings of robots and AI influence the audience's reception of performance works that feature such technologies and inspire artistic innovation. The chapters in Part Two, "Performing Arts Cultures and Robots/AI Developments," explore how theories and practices of the performing arts can engender critical dialogue on matters of cultural difference concerning culturally non-specific (though implicitly Western) framings of robotic and AI technologies within science and engineering contexts.

Reorienting the conversation around robotics and AI in the performing arts to place culture at its centre, *Cultural Technologies: Robots and Artificial Intelligence in the Performing Arts* offers thought-provoking analyses for advanced undergraduates, researchers, and performing arts practitioners interested in the relationships between music, theatre, and dance, and cutting-edge robotic and AI technologies.

Dr. Yuji Sone is a senior lecturer in the Department of Media, Communications, Creative Arts, Language and Literature at Macquarie University, Australia. His research has focused on the cross-disciplinary conditions of technologised performance.

Dr. Richard Savery is a developer of AI and robotics, using music and creativity as a medium to programme better interactions, understandings, and models. He is currently a research fellow at Macquarie University, Australia, developing new robotic musicians. His current research focuses on the creation of a new drumming and rapping robot, as well as robots painting to music and musical captcha.

CULTURAL TECHNOLOGIES

Robots and Artificial Intelligence in the Performing Arts

Edited by Yuji Sone and Richard Savery

Routledge
Taylor & Francis Group
NEW YORK AND LONDON

First published 2025
by Routledge
605 Third Avenue, New York, NY 10158

and by Routledge
4 Park Square, Milton Park, Abingdon, Oxon, OX14 4RN

Routledge is an imprint of the Taylor & Francis Group, an informa business

ISBN: 978-1-032-75859-6 (hbk)
ISBN: 978-1-032-75861-9 (pbk)
ISBN: 978-1-003-47597-2 (ebk)

DOI: 10.4324/9781003475972

Typeset in Sabon
by SPi Technologies India Pvt Ltd (Straive)

CONTENTS

ACKNOWLEDGEMENTS

The impetus for this book comes from two symposium events: "Japanese Culture and the Acceptance of Social Robots" in 2021 and "Robots, AI and Culture" in 2023, which were funded by TIFO (Toshiba International Foundation). These two symposia catalysed the editors' thinking on the juncture of robotic and AI technologies as they were framed in terms of cultural context. One of the editors of this book, Yuji Sone, organised the two events (with Naoko Abe, for the first event in 2021). Karl MacDorman presented a keynote talk at the 2021 seminar, which provided the basis for his chapter in this book. Richard Savery and Janet Biggs also presented a related panel discussion at the 2023 event. The editors would like to thank TIFO for their support for the seminar series, which led to the new and rich conversations that constitute this book.

Japanese roboticist Masahiro Mori passed away in January 2025. Mori is best known in the literature for his concept of the "uncanny valley." For the first seminar in 2021, Mori provided us with a short text (translated by Norri Kageki) that was read out at the opening of the event. In it, Mori discussed the importance of intuition for scientific discovery, and that intuition is nurtured through activity specific to one's cultural environment. This book is dedicated to Mori.

CONTRIBUTORS

Tristan Al-Haddad is the founder and CEO of Formations Studios, as well as a professor in the School of Architecture at the Georgia Institute of Technology. He holds a Master of Architecture from Georgia Tech and also studied at the University of Paris–La Villette and at the Daniels Center for Building Technology and Urban Design in Genoa, Italy.

Janet Biggs is a research-based interdisciplinary artist known for immersive work in video, film, and performance whose work focuses on individuals in extreme landscapes and situations, navigating the territory between art, science, and technology. Biggs has worked with institutions including NOAA, NASA, and CERN. Her work has garnered support and recognition from organisations such as the Guggenheim Foundation and the National Endowment for the Arts, *The New York Times*, *ArtForum*, and *ARTNews*, and has been exhibited and collected at museums and institutions worldwide. Biggs is a member of the New Museum's cultural incubator, NEW INC, and the Explorers Club.

Nicollete Cash is an audio producer and sound designer. Nicollette has a master's degree from Georgia Institute of Technology in Music Technology and received her undergraduate degree from George Mason University in Music Technology.

Emily S. Cross is based at ETH Zürich, where she leads the Professorship for Social Brain Sciences. She and her team explore questions concerning how we learn via observation, how motor expertise is manifest in the brain, the ever-evolving relationship between AI and art, and social influences on human-robot interaction. To address these questions, she uses intensive training procedures, functional neuroimaging, brain stimulation, and research paradigms that bridge technology, arts, and social sciences. At present, she and her team are particularly interested in how to build collaborations between the arts and AI, as well as how prolonged experience with robots changes how we perceive and interact with embodied artificial agents at brain and behavioural levels, and how these relationships manifest across the lifespan and in different cultures.

Denis Crowdy is an ethnomusicologist at Macquarie University. His research has explored Papua New Guinean popular music, music and mobile phones in the Pacific, and the role of code and software on musical activity. His output includes scholarly publications, creative works, and software.

Kohinoor M. Darda holds a PhD in Social/Cognitive Neuroscience from Bangor University, UK, and has previously worked as a postdoctoral researcher at the University of Glasgow (Glasgow, UK), Macquarie University (Sydney, Australia), and University of Pennsylvania (Philadelphia, PA, USA). She is currently the Founder and Director of ARISA Foundation, a research and social work organisation in Pune, India, that works at the intersection of neuroscience, psychology, and art to promote societal resilience and holistic well-being in children and adults. Kohinoor is also a trained Indian classical dancer (Bharatanatyam and Kuchipudi) and likes to integrate her background in the arts with her interest in neuroscience in her research and choreographic endeavours. Outside of research labs and dance studios, she enjoys social interactions with fictional characters in novels, underwater aesthetic experiences when scuba diving, and is often on the lookout for the next triathlon.

Evelyn Ficarra is a professor of Interdisciplinary Composition (Music) and co-director of the Centre for Research in Opera and Music (CROMT) at the University of Sussex. As composer, sound artist, and educator, Evelyn's work finds expression across a range of forms including music theatre, multimedia, installation, dance, film, and the concert hall. Primary research interests include investigations into sound objects in different contexts (theatre, installation): electroacoustic and multidisciplinary improvisation, and robot opera. Evelyn started the robot opera research project at CROMT in 2017, exploring issues of embodiment, performance, and vocality through the lens of radical technical innovation in opera.

Marina Hanganu is a theatre director and project manager specialising in telematic performance and creative technologies. She holds a PhD from UNATC "I.L. Caragiale" (Bucharest), an MA from the Royal Central School of Speech and Drama (London), and a BA in Theatre Directing from UNATC. Project initiator and artistic director of "Tele-Encounters" and "Tele-Encounters: Beyond the Human," co-funded by the Creative Europe programme of the EU. Author of the monograph (in Romanian) *Telematic Theatre* (Ed. Universitaria, 2022). Editor of *Beyond the Human: Theatre, Robots, and Social Realities* (Ed. Universitaria, 2023) and *Tele-Encounters: Telepresence and Migration* (UNATC Press, 2019).

Melissa Hill completed a Bachelor of Arts in Music Studies and a Bachelor of Science in Physics at Macquarie University. In her final year placements, she assisted in the study of human interactions with AI through the medium of a musical robot, as well as conducted research into undersea domain awareness, simulating responses to events detected using these systems and their applications to maritime safety. Melissa will commence her Master's degree in Wave Physics and Acoustics at Le Mans University in September 2024.

Tim Hopkins is a senior research fellow in Music at the University of Sussex and has a background in opera, music theatre and theatre direction, scenic design, moving image, and performing, in the UK and internationally. He is interested in opera and music theatre as a technological adventure and societal forum, researching new forms in performance, experimentation and collaboration, exploring how the artform responds to technologies and their meanings. His own work links historic and current practice, new spaces technology has created and the evolving artistic toolbox, to explore the impacts of technological transformation itself – cultural, emotional, political – and imagine possible futures for opera.

Elizabeth Jochum is an associate professor at the RELATE Research Laboratory for Art and Technology at Aalborg University. Her research uses the visual and performing arts as catalysts for re-thinking how we design and implement robots and other assistive technologies. From industrial robots to exoskeletons, Dr. Jochum's work involves transdisciplinary collaboration in human-robot interaction, health, and engineering to develop creative, value-sensitive, and human-centred approaches to ensure the technologies we build to address the real needs of people. She coordinates the ImprovAIze project on machine learning, dance improvisation, and wearable technologies.

Chris Kiefer is a senior lecturer in Music Technology, at Experimental Music Technologies Lab (Emute Lab) at the University of Sussex. Chris is a computer musician and musical instrument designer, specialising in musician-computer interaction, physical computing, and machine learning. He performs with custom-made instruments including malleable foam interfaces, touchscreen software, interactive sculptures, and a modified self-resonating cello. His work concentrates on machine learning and complex systems in new instrument design. He co-rans the AHRC Feedback Musicianship Network and runs the AHRC Musically Embodied Machine Learning project.

Amy LaViers works at the intersection of robotics and dance. Her writing, choreography, and machine designs have been presented internationally in both engineering and arts venues, including *Nature*, *Robotics and Autonomous Systems*, the MIT Press, Merce Cunningham's studios, the DANCE NOW Festival at Joe's Pub at the Public Theater, and the Performance Arcade. Her work has been featured in media outlets including *The New York Times*, *Forbes*, *Wired*, and *Dance Magazine*. She lives in Philadelphia where she directs the Robotics, Automation, and Dance (RAD) Lab, a non-profit for artmaking, commercialisation, education, outreach, and research.

James Leach is a social anthropologist who works on ecology and creativity based on research in Madang, Papua New Guinea, and with coders and contemporary artists in Europe and Australia. He has published extensively on art and intellectual property. James is a director of Research for the National Centre for Scientific Research (CNRS) in France.

Emily Liu is a computer science student at Georgia Institute of Technology. Emily has previous experience as a software developer with Amazon and eBay and has worked on computer vision processing with the robotic musicianship lab.

Karl F. MacDorman is the associate dean for Academic Affairs at the Luddy School of Informatics, Computing, and Engineering at Indiana University. He was previously an associate professor at the School of Engineering at Osaka University. He received his PhD in Computer Science from the University of Cambridge in 1997 and his BA in Computer Science from the University of California, Berkeley, in 1988. He has published more than 100 articles on human-robot interaction, machine learning, and cognitive science. His main research interests include the emergence and grounding of symbols, the development of cognitive robotics, android science, and the uncanny valley.

Trinity Melder is a research student at Macquarie University, Australia, studying human-computer interactions. With a master's in research and a background in computer science, Trinity focuses on enhancing the interactions of the human-computer experiences through innovative technologies.

Hope Phan is currently a Cloud engineer for Oracle. Hope graduated from Georgia Institute of Technology in 2024 with a BS in Computer Science. At Georgia Tech, Hope was also a Dancer in a variety of performances, including performances with medusai.

Jiahe Qian is an HCI prototype engineer for Shure's AI team. Jiahe received his undergraduate degree in Music Technology from Georgia Institute of Technology and is currently receiving his Master's in Music Technology.

Tanvi Raghuram has a master's degree in clinical Neuropsychology from Bangor University, UK, and a triple major bachelor's degree in communication and media studies, English Literature, and Psychology from CHRIST (Deemed to be University), Bangalore. She is currently working as a research assistant at the ARISA Foundation in Pune. Tanvi Identifies as an intersectional feminist whose work aligns with her identity. As a research enthusiast, she has worked in the fields of cognitive and developmental psychology, clinical neuropsychology, and cultural media literacy – focusing on topics such as coping styles, self-body representations, and female representation in mainstream media, respectively. She was also a student researcher at the Cognitive, Affect, and Behaviour Lab in CHRIST (Deemed to be University) and worked on the Government of India, Department of Science and Technology funded cross-sectional project. Outside of research, Tanvi is a classical dancer with an interest in literary discourse and poetry, and an active advocate for sustainable eco-friendly living.

Amit Rogel is a PhD researcher in Robotic Musicianship at Georgia Tech Center for Music Technology. He has a background in both music (Trombone and Production) and mechanical engineering (Robotics and speaker design). He received his bachelor's in mechanical engineering from the Rochester Institute of Technology. His current work focuses on the correlation between sound and robotics in two ways: using robots to create gestural, expressive responses to music and creating musical responses to improve safety and fluency in human-robot interaction. Amit also works with robotic musicians and dancers, with whom he has performed internationally.

Richard Savery is a developer of AI and robotics, using music and creativity as a medium to programme better interactions, understandings, and models. His research has received $3.3 million in funding and significant industry partnerships. His work has been featured in *Scientific American*, BBC Radio, and Disney +. He completed a PhD in Music Technology (minor in Human-Computer Interaction) at the Georgia Institute of Technology, graduated in 2021.

Hannah Schlisky is an Unreal Engine technical director and virtual production engineer at Lux Machina. She graduated from Alfred University in Electronic Integrated Arts and Sculptures.

Yaron Shyldkrot is a practitioner-researcher and performance maker. His research and artistic work explore the blurriness between natural and artificial relationships with technology and each other, and the edges of sound and vision. Yaron is a lecturer in Scenography and Design-led Performance at the University of Leeds, UK and co-convenor of the Intermediality in Theatre and Performance Research Working Group at the International Federation for Theatre Research (IFTR). His monograph, *Performing Robots and AI Narratives: Misbehaving Machines*, is forthcoming with Bloomsbury.

Yuji Sone is a senior lecturer in the Department of Media, Communications, Creative Arts, Language, and Literature at Macquarie University in Australia. Yuji is a performance researcher who initially trained with the experimental theatre company Banyu-Inryoku in Japan. He teaches performing arts and creative industries courses with an emphasis on praxis in relation to cultural studies and media studies. His research has focused on the cross-disciplinary conditions of technologised performance. He is the author of *Japanese Robot Culture: Performance, Imagination, and Modernity* (2017), which examines Japanese affinity for the robot through contemporary performing robots.

Ripken Walker is a project engineer at For Audio Video and founder of RAW mobile studio. Ripken graduated with his undergraduate and master's in music technology from Georgia Institute of Technology.

Gil Weinberg is a professor and the founding director of the Georgia Tech Center for Music Technology, where he leads the Robotic Musicianship group. His research focuses on developing artificial creativity and musical expression for robots and augmented humans. Weinberg presented his work worldwide in venues such as the Kennedy Center, the World Economic Forum, Ars Electronica, the Smithsonian Cooper-Hewitt Museum, SIGGRAPH, TED-Ed, DLD, and others. He received his MS and PhD degrees in Media Arts and Sciences from MIT and his BA from the interdisciplinary programme for fostering excellence at Tel Aviv University.

PART ONE

Robot/AI Cultures and Performing Arts Practices

1

INTRODUCTION

New Conversations across the Performing Arts

Yuji Sone

This book explores the impacts of robotic and AI technologies on performance practices, highlighting the cultural contexts for the respective languages of music, dance, and theatre. When we speak of culture in this book, we are concerned with a culture's accumulated shared knowledge and stored information that binds people together within that culture and potentially beyond the confines of nationality. The aim of this book is to investigate how robotic and AI technologies, when they are used in music, dance, and theatre, shape or influence existing performing arts languages in a given culture. We are also interested in the other side of this question: How do music, dance, and theatre shape the translation, transference, and adoption of robotic and AI technologies? According to literary scholar Stephen Greenblatt, "creative *metissage*," a French term meaning syncretic processes, mixing, and hybridity arising from cultural and artistic encounters, ignores persistent forces for protecting boundaries and cultural impulses towards fixity (2009, 1). The varied discussions in this book's chapters likewise emphasise how "texts, images, artifacts, and ideas are moved, disguised, translated, transformed, adapted, and reimagined" (Greenblatt et al. 2009, 4). In fact, robotic and AI technologies, which, from some science and engineering perspectives, might see themselves as culturally agnostic products of science, not only bring attitudes, assumptions, and even embodiments across cultural boundaries but act as agents or catalysts for further productive cultural "metissage." Starting from this positive argument for the mixed results of cultural encounters, the contributors in this book, a mix of theorists, practice-based researchers, and an artist examine the contexts and occurrences of specific works as case

DOI: 10.4324/9781003475972-2

studies to illuminate the way that robots and AI are understood in a culture, as expressed in terms of the performing arts. As such, the book focuses on some key modalities of the performing arts (theatre, music, dance, and opera), placing them at the centre of discussions about robotics and AI in relation to culture rather than beginning with the technologies themselves, as in the dominant debates on robotics and AI. Part of our enquiry in this book involves considering how (sub)cultures of artistic practices actively and critically engage with robotics and AI.

Some Key Concepts in This Book

It is useful at this point in our framing of this book to consider "cultural technologies." Media and communication scholar Göran Bolin's *Cultural Technologies: The Shaping of Culture in Media and Society* (2012) discusses emerging media technologies in "cultures of technology" in which daily lives are infused with technologies (2012, 2). For Bolin, media cultural technologies are the tools that allow one to select, store, or edit the information "by which culture is formed, reproduced or changed" while culture and technology interact with each other in structuring and being structured (2012, 8). In this book, we keep such dialectic forces within the structured formats of the performing arts in mind, focusing on emerging technologies of robotics and AI. We investigate interactions between cultural languages of the performing arts, i.e., performing arts iteratively shaped by and shaping a given culture, and artistic adaptations of robotics and AI, which also involve iterative modes of reception and sharing of forms and modalities.

As we have indicated above, this book aims to bring perspectives from the performing arts "cultures" as they encounter technology forward, rather than start with the technology, per se. Having said that, it is necessary to indicate that analyses of these technologies have not ignored culture, the interface between technology and the arts, or technology and creativity. Researchers outside the creative arts in more recent years have been examining notions of "creativity" in a broader context. From scientific perspectives, several studies have analysed the relationship between creativity and technology, such as Du Sautoy (2019), Veale and Cardoso (2019), Miller (2014, 2019), and Varela (2021). As AI technologies have become very sophisticated, arts practices utilising technology can be regarded as a field of research for varied disciplines. In *The Creativity Code: Art and Innovation in the Age of AI* (2019), mathematician Marcus Du Sautoy discusses AI through a frame of relationality between humans and machines, positing art and mathematics as games that require creative intelligence. Computer scientists Tony Veale and F. Amílcar Cardoso posit their study of creativity as a branch of scientific study of artificial intelligence in *Computational Creativity: The Philosophy and Engineering of Autonomously Creative Systems* (2019). Narrowing the focus of his previous book *Colliding Worlds: How Cutting-Edge Science Is Redefining Contemporary Art* (2014), which discusses how aesthetic works can be produced at the nexus of science and art, in his book *The Artist in the Machine: The World of AI-Powered Creativity* (2019), science historian Arthur Miller surveys creative endeavours with AI technologies including practices in visual art, music, text, and musical theatre, and looks at positive possibilities in creative use of such technologies. Web developer Miguel Escobar Varela's *Theater as Data: Computational Journeys into Theater Research* applies uses of computational methods, such as statistical analysis and data visualisation, to theatre disciplines, highlighting tensions in such interdisciplinary approaches of the digital humanities (2021). The chapters in this book, on the other hand, deploy specific contextual frames for current practices and related discussions on the use of AI and robotic technologies within selected disciplines of the performing arts.

This book also moves away from broader views of "art" and its intrinsic relationships with technology within wider creative contexts in humanities as in existing studies such as McCormack and D'Inverno (2012), Broadhurst and Price (2017), Zylinska (2020), Audry (2021), Bown (2021), or Filimowicz (2023). Before the current AI boom, the combined themes of computer, art, music, and creativity were already explored in *Computers and Creativity* (McCormack and D'Inverno 2012) by artists and researchers in computing Jon McCormack and Mark d'Inverno. Performance studies scholar Susan Broadhurst and researcher in Human-Computer Interaction (HCI) Sara Price's *Digital Bodies: Creativity and Technology in the Arts and Humanities* (2017) explores possibilities of creative engagement with technology, covering diverse themes such as the performing body, HCI, fashion, museum, and medical simulation. Similarly, a positive vision for the use of AI in the visual arts, installation art, and performance art is highlighted in *Art in the Age of Machine Learning* (2021) by transdisciplinary artist-researcher Sofian Audry. Music video maker and researcher Michael Filimowicz's *AI and the Future of Creative Work: Algorithms and Society* (2023) discusses the emerging hybridised nature of creative practices and industry practices with computer and digital technologies. From a practitioner's point of view, musician and musicologist Ollie Bown discusses the impacts of automation and computational creativity through generative technologies concerning art and music from multidisciplinary perspectives, including computer science, psychology, and creative art research (2021). Media artist and scholar on new media Joanna Zylinska, on the other hand, engages with "AI art" with the view that human creativity is "partly computational," critiquing corporate technology-driven visual artworks as foreclosing critical engagement (2020, 54). The focus of our book is deliberately narrower than these existing book-length publications on artistic creativity and cutting-edge technologies of robotics and AI, seeking to highlight cultural inflections in these performing arts practices with robotic and AI technologies.

Focusing on technology-based artworks in performing arts settings, the current book differs from books about interactive art, new media art, or performance art that utilise sophisticated media technologies with algorithms and automated systems. These forms have been examined in Leonardo publications (Cubitt and Thomas 2013; Rinehart and Ippolito 2014; Broeckmann 2016; Penny 2017). A similar approach is deployed in *Robots and Art: Exploring an Unlikely Symbiosis* (Herath, Kroos, and Stelarc 2016), which examines contemporary robotic art, reflecting viewpoints of artists, engineers, and scientists. As well, sound art and the visual arts are discussed in *Cultural Robotics: Social Robots and Their Emergent Cultural Ecologies* (Dunstan et al. 2023). With an emphasis on social robotics, the evolving technological environments for sound art and the visual arts are discussed as "cultural ecologies" that include practices outside the creative arts disciplines. While sound is the key medium of enquiry in *Sound and Robotics: Speech, Non-verbal Audio and Robotic Musicianship* (2024) by Richard Savery, its focus is on "sound out from the robots" in their interaction with humans (2024, 2). These publications highlight strategic arguments for technology-based arts' intersection with disciplines of HCI or HRI (Human-Robot Interaction), interdisciplinary research fields, combining enquiries of philosophy, anthropology, sociology, psychology, cognitive sciences, biology, or ethology. This collection, in contrast, brings in authors who deploy tools of musicology, theatre studies, and dance studies, while referring to other disciplines, to examine works within the frames of the performing arts that separate the audience from the performer on the stage.

Rather than broadly situating creativity and contemporary technologies in the contexts of media, science, philosophy of art, media art, or the visual arts, this book stresses the viewpoints of practitioners in the performing arts. The case studies in this volume examine the nature of events that interact with robotics and AI technologies within performance frames that foreground their relevant disciplinary language(s). In other words, the authors address the challenges faced by the works' creators that are specific to the performing arts when a work engages with robotics and AI technologies. Therefore, critical questions such as these are necessarily considered:

- Debates on robotics and AI have tended either to celebrate or condemn the advancement of robotic and AI technologies, eliciting responses such as hope or fear. Can discussions in music, dance, or theatre avoid such gross generalisations if they choose to focus on these technologies *and* the cultural context in which such technologies are presented?
- On the other hand, how do the practices of a particular cultural context inform the use of robots and AI in the music, dance, and theatre of that culture? And, relatedly, what is at stake in the context of globalisation?
- Does the use of AI in music, dance, and theatre generate a reification of these art practices in a particular cultural context? For example, in the case of music, AI treats music as an end product, as opposed to musicking, referring to the more common and historically past idea that music is a process. Similarly, does AI only treat dance in terms of obvious patterns and forms rather than recognising a particular dance's meanings in the context of culture?
- When the audience views robots on the stage, they may change their reactions according to presented narratives that contextualise these robots. Is it possible to see such adaptive changes as catalysing new understandings in the appreciation of performance by an audience that is witnessing the work precisely at the meeting place of the "language" for that artistic discipline and newer technologies' contribution?

In the course of each chapter's enquiry, this book highlights nuanced differences and the closeness between some performing arts practices when they are working with robotics and AI. This is another area that existing studies of music, theatre, and dance that utilise robotics and AI and have not explored: the interconnections and differences between performing arts genres in relation to their use of robotics and AI. The authors present their discussions in styles that are aligned with the standards and expectations of their own disciplines. *Cultural Technologies: Robots and Artificial Intelligence in the Performing Arts* offers a starting point for such conversations across the performing arts disciplines.

It is necessary to first discuss different paths that have been taken by scholars of music, theatre, and dance studies on themes of robotics and AI.

Music has had a long engagement with technology. For example, music theorist James Perone traces back the history of the relationship between music and technology to Pythagoras of ancient Greece and his tuning system, and then chronologically discusses historical technological developments of musical instruments, notation, and recording (Perone 2022). The contributors to *Music and Technology in the Twentieth Century* (Braun 2002) discuss notable developments in music in the 20th century such as the 45 RPM record, the tape recorder, and the synthesiser. Mark Katz, a scholar on music and technology, focuses on the significant impacts of recording technologies in his book *Capturing Sound: How Technology Has Changed Music* (2010). While previously the development of digital technologies and their effects on music attracted focused scholarly attention (Hugill 2012; Cresswell-Jones and Bennett 2015), research in the 2020s on music has concerned itself with artificial intelligence and robotics. For example, the authors in *Handbook of Artificial Intelligence for Music: Foundations, Advanced Approaches, and Developments for Creativity* (Miranda 2021) and *Artificial Intelligence and Music Ecosystem* (Clancy 2023) discuss the potentialities and implications of ever-developing AI technologies in music. With robotics, researchers in *Musical Robots and Interactive Multimodal Systems* (Solis and Ng 2011) focus on live performance and machine performers' interactions that are presented to the audience.

However, these studies above take for granted that technology is culturally neutral. Timothy Dean Taylor in *Strange Sounds: Music, Technology and Culture* (2001) includes, in addition to a discussion of exploratory practices with new technologies, a critical analysis of the implications of digital technologies, seeing them as negatively affecting music globally. This book not only updates critical discussions such as Taylor's by focusing on robotic and AI technologies but also explores underdeveloped discussions on the combination of music and robots as an approach that is inviting the emergence of new cultural products.

While there is an area of academic enquiry on theatre technologies, that is, relating to theatre architecture, scenography, and stage design, including puppetry and automation that examines historical and contemporary cases (Baugh 2013; Butterworth and Normington 2017; Oliszewski, Fine, and Roth 2018; Nawata and Dethlefs 2022), the main thrust for theatre and dance studies scholars in their recent scholarship on theatre technologies is to do with investigations of innovative uses of modern and contemporary technologies on stage. In relation to the historical avant-garde of the early 20th century, numerous book-length publications in theatre and performance studies in the new millennium discuss the development of theatre performance practices taking up digital media technologies (Giannachi 2004; Smith 2005; Berghaus 2005; Chapple and Kattenbelt 2006; Broadhurst and Machon 2006; Dixon 2007; Matthew Causey 2006; Giannachi 2007; Auslander and Auslander 2008; Birringer 2008; M. Causey, Meehan, and O'Dwyer 2015; Broadhurst and Price 2017; Broadhurst 2009; Salter 2010; Broadhurst and Machon 2009; Parker-Starbuck 2011; Klich and Scheer 2012). As the human body is often the focus of theatre and dance studies analyses, scholars discuss the intertwinement of the body with technologies. Scholars have used terms such as "liveness," "cyborg," "posthuman," "virtuality," "argumentation," or "interactivity" to explain their investigations. In dance, on the other hand, there is a strong tendency to focus on the material human body and its movements in digital environments or in the context of HRI (Kozel 2007; Rosenberg 2012; Bleeker 2017; LaViers and Maguire 2023). The latest additions to the list of studies of theatre, performance, and dance are *Deviations in Contemporary Theatrical Anthropology: New Myths and Performative Rituals between XR, Robots and AI* (Fuoco 2024), *The Routledge Companion to Performance & Technology* (Bleeker and Shaw, forthcoming), *Revealing Posthuman Encounters in Performance* (Boselli and Lucie, forthcoming), and *Performing Robots and AI Narratives: Misbehaving Machines* (Shyldkrot, forthcoming), which update the debates in theatre, dance, and performance studies, including discussions on current uses of robotics and AI in performance practices. Just as for the discipline of music, cultural influences upon and the implications of the artistic use of robotic or AI technologies in works of theatre, dance, or performance are rarely the focus of academic enquiry and investigation.

Exploring a new approach that highlights aspects of culture in discussion of contemporary performing arts practices that engage with robotic and AI technologies, the chapters in this book are organised into two sections. In the chapters included in the first section, "Part One: Robot/AI Cultures and Performing Arts Practices," contributors discuss how cultural understandings of robots and AI could influence the audience's reception of performance works that feature such technologies as well as inspire artistic explorations in performance. In "Part Two: Performing Arts Cultures and Robots/AI Developments," the chapters explore how theories and practices of the performing arts can contribute to, or could be used as a tool for, critical dialogue with often culturally non-specific (though implicitly Western) discussions of developments in robotic and AI technologies in science and engineering contexts.

In Chapter 2, as the initial discussion of Part One, Yaron Shyldkrot explores the creative potential of "glitches" in robotic performance. Western popular narratives of humanoids and androids usually highlight the differences between humans and anthropomorphic machines, such that the intertwined emotions of fear towards robots and the human desire to control them emerge through conflict. By analysing Rimini Protokoll's performance works *Remote X* (2013) and *Uncanny Valley* (2018), Shyldkrot challenges established views of human-robot interactions in Western narratives. He highlights how deliberate technological imperfections disrupt assumptions around "flawless" robots, opening up new dramaturgical spaces where AI's limitations become generative rather than restrictive. Introducing the theme of "malfunction" as a source of creative inspiration facilitates the reimagining of performance, opening new possibilities when robots deviate from expected norms.

The following two chapters present analyses of practice-based projects. In Chapter 3, Amit Rogel et al. discuss the "medusai" project under Gil Weinberg, the director of the Center for Music Technology at the Georgia Institute of Technology in the US. The project features a large-scale robotic sculpture that resembles the head of Medusa. It is a performance installation project in which uneasiness towards the figure of the robot is fused with affect relating to the myth of Medusa – a figure embodying both monstrosity and fascination in the West. In its multimodal performance presentations, the sculpture's robotic arms pluck strings that are attached to the structure, generating melodies. As well, the work's dance and music performers are able to pluck the strings to generate sound, while the machine responds through sound, light, and movement. Rogel et al. discuss how the project explores the dancers' and musicians' reactions of uncertainty and curiosity, as they then *adapt* their performances when interacting with the robotic installation.

Marina Hanganu in Chapter 4 shifts focus to the ethical and cultural dimensions of robots in caregiving roles, specifically in European contexts. The telematic theatre performance *2032 SMART-FAMILY* (2022) utilises a cartoonish, cuddly image of a friendly assistive robot for audiences in Italy and Romania. Hanganu examines how social robots can mediate transnational family dynamics and caregiving, particularly among Romanian migrants in Italy. *2032 SMART-FAMILY* underscores how robots based in community and family contexts of the future could both support and complicate cross-border family practices, offering a culturally nuanced view of care and automation.

Turning to the impact of AI on dance aesthetics, Chapter 5, by Tanvi Raghuram et al., investigates one role of AI in choreography and dance. The authors, a collaborative research team whose members come from the Social Brain Sciences Group at the Federal Institute of Technology, Zurich, Switzerland, and the Advancement and Research in the Sciences and Arts Foundation (ARISA) in Pune, India, discuss how AI-generated choreography shapes aesthetic preferences and audience responses. The authors explore the emotional impacts of cultural dispositions in collaboration between human dancers and AI performers. This chapter introduces a new way of evaluating dance by considering the broader cultural and cognitive implications of AI in dance, and how these technologies could potentially influence the evolution of dance's disciplinary language(s).

Concluding the first section, Chapter 6 by Yuji Sone focuses on artist Keiichiro Shibuya's android opera – a combination of advanced robotics and operatic performance. Shibuya employs the androids "Skeleton" and "Alter," designed by noted Japanese engineers Hiroshi Ishiguro and Takashi Ikegami, as central performers in Shibuya's operas *Scary Beauty* (2017) and *Mirror* (2022). These androids serve as reciters, singers, and conductors alongside live orchestras in works that explore themes of machinic otherness and the end of humanity. Sone critically examines Shibuya's dramaturgical strategies, highlighting how his work navigates cultural biases to shape reception in both Japan and the West. Sone explains that the works deploy a type of orientalism within the Western tradition of the opera and utilise the robot's "otherness" in order to leverage Japanese cultural expectations.

In Part Two, "Performing Arts Cultures and Robots/AI Developments," contributors consider how works of music, theatre, or dance could provide productive yet critical frames on AI technologies in a given culture, highlighting translation, transference, and adoption.

Part Two begins with Chapter 7, in which Karl MacDorman utilises historical Japanese aesthetic theories of *ningyō jōruri* (Japanese traditional puppet theatre with storytelling and music, also known as Bunraku) with regard to the design of humanoid robots. MacDorman proposes relations between the theories of 17th- and 18th-century Japanese dramatist CHIKAMATSU Monzaemon and MORI Masahiro, a contemporary Japanese roboticist who became known for his discussion of the "uncanny valley," offering insights into the design of anthropomorphic robots that balances human likeness with aspects of non-representational cultural resonance. Building on his past research concerning the phenomenon of the uncanny valley, MacDorman explores parallels between the traditional Japanese theory of puppetry and contemporary robotics.

In Chapter 8, Elizabeth Jochum et al. discuss how collaboration in an operatic work is transformative and can envelop participants, including people with non-art-related backgrounds, such as scientists and engineers. They refer to the ongoing *Robot Opera* (2017–) research project at the Centre for Research in Opera and Music Theatre (CROMT) at the University of Sussex in England, which explores machine performers' potentiality when integrated with musical improvisation and puppetry. Jochum et al. also reveal the design restrictions of commercial robots when used in performance and highlight engineers' limited experiences with the specific requirements of artistic creations. This chapter reflects on the broader implications of the presence of robots and AI on stage and how they provoke questions of "authentic" voice. This chapter considers how machines can engage in creative processes, traditionally the domain of human performers in the West, and how performing arts practices such as opera can drive new technological innovation.

The following two chapters in Part Two bring critical perspectives to applications of robotics and AI technologies with regard to dance and music. Chapter 9 by Amy LaViers adds a critical dimension to the book's enquiry into the limits of robotic embodiment in performance, emphasising the importance of recognising cultural specificity when analysing movement and expression. LaViers presents the concept of a "body-language dictionary" AI tool for cataloguing human movement. Focusing on human bodily movement and gesture, LaViers explores how algorithms may be incompatible with cultural practices. She argues that, while AI and robotics might mimic human gestures and posture, they cannot fully capture the cultural or contextual meaning of bodily expressions.

In Chapter 10, Denis Crowdy and James Leach examine the impact of AI on music, particularly concerning voice, authorship, and ownership in the West. It offers a critical perspective on the protection of intellectual property rights over vocal works, referring to the Human Artistry Campaign (HAC). This campaign was launched by key music industry groups in the US, including the Recording Industry Association of America (RIAA) and the Recording Academy, at the South by Southwest (SXSW) Conference in Texas in 2023. AI technologies such as the audio deepfake are believed to exacerbate current issues regarding commercial notions of property and copyright for voice recording in music. HAC is concerned that, by dividing sound from its source, AI-generated voice disregards the unique and irreplaceable indexicality of human identity and embodiment. The authors explore such points within broader debates about the relationship between a human source, vocal sound, and place with reference to cultural music practices in Papua New Guinea, and in relation to conceptions of voice in HRI that function outside the established framework concerning intellectual property.

The book concludes with two chapters that discuss the wider creative potential of robotics and AI technologies in the performance of music, dance, and vocal works. Chapter 11, by Richard Savery et al., focuses on *Keirzo*, a drumming and rapping robot, and its public performances during 2023 at the Parramatta Lanes Festival in Parramatta, a satellite city of Sydney, Australia. Savery examines adults' and children's preconceptions and predispositions regarding robots and rap music, and how different demographics in Australia engaged with a "creative" robot musician. This chapter analyses how robotic performers can shape human-robot interactions in public settings, discussing how rap music culture may act as a "social lubricant" for robot-human interaction. Savery's consideration of the role of cultural practices in the context of a festival may indicate ways in which the acceptance of robots in everyday life could be facilitated or hindered. Finally, in Chapter 12, writing from a deeply personal perspective, artist Janet Biggs presents an artist's account of her work *A.I. Anne* (2019–), which involves a photographic image of her late aunt that reacts to vocal and sound sources. Biggs emphasises the ethical and emotional dimensions of working with AI as a collaborator in artistic practice. Drawing upon the communication methods employed by her aunt, who lived with autism, the project demonstrates the potential of AI to foster empathy and encourage inclusivity in the arts. By discussing multimedia performances, including a livestreamed event that was held during the COVID-19 pandemic in 2020 and her performance of *A.I. Anne* at the United Nations Headquarters in 2023, Biggs showcases how AI can effectively further social advocacy by bridging art and science, especially with regard to underrepresented communities. She concludes that while this technology's reach and ability to effect change is vast, there is power in one person's story: by emphasising the human element within technological advancements, it may be possible to catalyse social change.

References

Audry, Sofian. 2021. *Art in the Age of Machine Learning*. The MIT Press. 10.7551/mitpress/12832.001.0001

Auslander, Philip, and Philip Auslander. 2008. *Liveness*. Routledge. 10.4324/9780203938133

Baugh, Christopher. 2013. *Theatre, Performance and Technology: The Development and Transformation of Scenography*. Second edition. Theatre and Performance Practices. Basingstoke: Palgrave Macmillan.

Berghaus, Günter. 2005. *Avant-Garde Performance: Live Events and Electronic Technologies*. Houndmills, Basingstoke, Hampshire; New York: Palgrave Macmillan.

Birringer, Johannes H. 2008. *Performance, Technology, & Science*. New York; New York: PAJ Publications.

Bleeker, Maaike. 2017. *Transmission in Motion: The Technologizing of Dance*. London: Routledge.

Bleeker, Maaike, and Norah Zuniga Shaw, eds., forthcoming. *Routledge Companion to Performance & Technology*. New York: Routledge.

Bolin, Göran, ed. 2012. *Cultural Technologies: The Shaping of Culture in Media and Society*. Routledge Research in Cultural and Media Studies. New York: Routledge.

Boselli, Stefano, and Sarah Lucie, eds., forthcoming. *Revealing Posthuman Encounters in Performance*. New York: Routledge.

Bown, Ollie. 2021. *Beyond the Creative Species: Making Machines That Make Art and Music*. Cambridge, Massachusetts: The MIT Press.

Braun, Hans-Joachim. 2002. *Music and Technology in the Twentieth Century*. John Hopkins ed. Baltimore, Md: John Hopkins University Press.

Broadhurst, Susan. 2009. *Digital Practices: Aesthetic and Neuroesthetic Approaches to Performance and Technology*. Houndmills; New York: Palgrave Macmillan.

Broadhurst, Susan, and Josephine Machon, eds. 2006. *Performance and Technology*. London: Palgrave Macmillan UK. 10.1057/9780230288157

Broadhurst, Susan, and Josephine Machon, eds. 2009. *Sensualities/Textualities and Technologies*. London: Palgrave Macmillan UK. 10.1057/9780230248533

Broadhurst, Susan, and Sara Price, eds. 2017. *Digital Bodies: Creativity and Technology in the Arts and Humanities*. 1st ed. 2017. Palgrave Studies in Performance and Technology. London: Palgrave Macmillan UK. 10.1057/978-1-349-95241-0

Broeckmann, Andreas. 2016. *Machine Art in the Twentieth Century*. Leonardo Book Series. Cambridge, MA: The MIT Press.

Butterworth, Philip, and Katie Normington, eds. 2017. "Medieval Theatre Performance: Actors, Dancers, Automata and Their Audiences." In *Medieval Theatre Performance*. Boydell and Brewer. 10.1515/9781787440784

Causey, M., E. Meehan, and N. O'Dwyer. 2015. *The Performing Subject in the Space of Technology: Through the Virtual, Towards the Real*. 1st ed. 2015. Palgrave Studies in Performance and Technology. London: Palgrave Macmillan. 10.1057/9781137438164

Causey, Matthew. 2006. *Theatre and Performance in Digital Culture from Simulation to Embeddedness*. London; New York: Routledge.

Chapple, Freda, and Chiel Kattenbelt. 2006. *Intermediality in Theatre and Performance*. Amsterdam: Rodopi.

Clancy, Martin, ed. 2023. *Artificial Intelligence and Music Ecosystem*. Abingdon, Oxon: Routledge.

Cresswell-Jones, Angela, and Rebecca Jane Bennett, eds. 2015. *The Digital Evolution of Live Music*. First edition. Chandos Information Professional Series. Waltham, MA: Chandos Publishing.

Cubitt, Sean, and Paul Thomas, eds. 2013. *Relive: Media Art Histories*. Cambridge, MA: The MIT Press.

Dixon, Steve. 2007. *Digital Performance: A History of New Media in Theater, Dance, Performance Art, and Installation*. Cambridge, Mass.: MIT Press.

Du Sautoy, Marcus. 2019. *The Creativity Code: Art and Innovation in the Age of AI*. Cambridge, Massachusetts: The Belknap Press of Harvard University Press.

Dunstan, Belinda J., Jeffrey T. K. V. Koh, Deborah Turnbull Tillman, and Scott Andrew Brown. 2023. *Cultural Robotics: Social Robots and Their Emergent Cultural Ecologies*. 1st ed. Cham: Springer International Publishing AG.

Escobar Varela, Miguel. 2021. *Theater as Data: Computational Journeys into Theater Research*. Ann Arbor: University of Michigan Press. 10.3998/mpub.11667458

Filimowicz, Michael. 2023. *AI and the Future of Creative Work: Algorithms and Society*. Abingdon, England: Routledge.

Fuoco, Ester. 2024. *Deviations in Contemporary Theatrical Anthropology: New Myths and Performative Rituals between XR, Robots and AI*. London: Routledge. 10.4324/9781032676937

Giannachi, Gabriella. 2004. *Virtual Theatres: An Introduction.* London; New York: Routledge.

Giannachi, Gabriella. 2007. *The Politics of New Media Theatre: Life TM.* London and New York: Routledge.

Greenblatt, Stephen, Ines Županov, Reinhard Meyer-Kalkus, Heike Paul, Pál Nyíri, and Frederike Pannewick. 2009. *Cultural Mobility: A Manifesto.* 1st ed. Cambridge: University Press. 10.1017/CBO9780511804663

Herath, Damith, Christian Kroos, and Stelarc, eds. 2016. *Robots and Art: Exploring an Unlikely Symbiosis.* Cognitive Science and Technology. Singapore: Springer Singapore Pte. Limited. 10.1007/978-981-10-0321-9

Hugill, Andrew. 2012. *The Digital Musician.* Second edition. London: Routledge. 10.4324/9780203111796

Katz, Mark. 2010. *Capturing Sound: How Technology Has Changed Music.* Revised edition. Acls Humanities E-Book. Berkeley: University of California Press. 10.1525/9780520947351

Klich, Rosemary, and Edward Scheer. 2012. *Multimedia Performance.* Houndmills, Basingstoke, Hampshire; New York, NY: Palgrave Macmillan.

Kozel, Susan. 2007. *Closer: Performance, Technologies, Phenomenology.* Leonardo (Series) (Cambridge, Mass.). Cambridge, Mass: MIT Press.

LaViers, Amy, and Catherine Maguire. 2023. *Making Meaning with Machines: Somatic Strategies, Choreographic Technologies, and Notational Abstractions through a Laban/Bartenieff Lens.* The MIT Press. 10.7551/mitpress/14060.001.0001

McCormack, Jon, and Mark D'Inverno. 2012. *Computers and Creativity.* Berlin: Springer-Verlag.

Miller, Arthur I. 2014. *Colliding Worlds: How Cutting-Edge Science Is Redefining Contemporary Art.* 1st edition. New York: *Norton Agency Titles.

Miller, Arthur I. 2019. *The Artist in the Machine: The World of AI-Powered Creativity.* Cambridge, Massachusetts: MIT Press.

Miranda, Eduardo Reck, ed. 2021. *Handbook of Artificial Intelligence for Music: Foundations, Advanced Approaches, and Developments for Creativity.* 1st Edition 2021. Cham: Springer International Publishing AG. 10.1007/978-3-030-72116-9

Nawata, Yuji, and Hans Joachim Dethlefs, eds. 2022. *Performance Spaces and Stage Technologies A Comparative Perspective on Theatre History.* 1st ed. Theater 146. Bielefeld: transcript Verlag. 10.1515/9783839461129

Oliszewski, Alex, Daniel Fine, and Daniel Roth. 2018. *Digital Media, Projection Design & Technology for Theatre.* 1st edition. Oxford: Routledge. 10.4324/9781315666976

Parker-Starbuck, Jennifer. 2011. *Cyborg Theatre: Corporeal/Technological Intersections in Multimedia Performance.* Houndmills, Basingstoke, Hampshire; New York: Palgrave Macmillan.

Penny, Simon. 2017. *Making Sense: Cognition, Computing, Art, and Embodiment.* Cambridge, Massachusetts London: The MIT Press.

Perone, James E. 2022. *Music and Technology: A Historical Encyclopedia.* 1st ed. Santa Barbara: Greenwood.

Rinehart, Richard, and Jon Ippolito. 2014. *Re-Collection: Art, New Media, and Social Memory*. The MIT Press. 10.7551/mitpress/7975.001.0001

Rosenberg, Douglas. 2012. *Screendance: Inscribing the Ephemeral Image*. 1st ed. New York: Oxford University Press. 10.1093/acprof:oso/9780199772612.001.0001

Salter, Chris. 2010. *Entangled: Technology and the Transformation of Performance*. Cambridge, Mass.: MIT Press.

Savery, Richard, ed. 2024. *Sound and Robotics: Speech, Non-Verbal Audio and Robotic Musicianship*. Boca Raton, FL: CRC Press.

Shyldkrot, Yaron., forthcoming. *Performing Robots and AI Narratives: Misbehaving Machines*. London: Bloomsbury.

Smith, Marquard, ed. 2005. *Stelarc: The Monograph*. Cambridge, Mass.; London: MIT Press.

Solis, Jorge, and Kia Ng, eds. 2011. *Musical Robots and Interactive Multimodal Systems*. Springer Tracts in Advanced Robotics, 74. Berlin, Heidelberg: Springer Berlin Heidelberg. 10.1007/978-3-642-22291-7

Taylor, Timothy Dean. 2001. *Strange Sounds Music, Technology & Culture*. New York: Routledge.

Veale, Tony, and F. Amílcar Cardoso, eds. 2019. *Computational Creativity The Philosophy and Engineering of Autonomously Creative Systems*. 1st ed. 2019. Computational Synthesis and Creative Systems. Cham: Springer International Publishing. 10.1007/978-3-319-43610-4

Zylinska, Joanna. 2020. *AI Art: Machine Visions and Warped Dreams*. Media: Art: Write: Now. London: Open Humanities Press.

2

I AM GOING TO TURN INTO PETER

Glitching Human-Robot Interactions

Yaron Shyldkrot

Introduction

This chapter examines human-robot interactions (HRI) through the incorporation of "malfunctioning" humanoids in performance events. Within this volume's study of the interrelationships between robotic and artificial intelligence (AI) technologies and performance practices, I respond to the notable – predominantly Western – cultural languages and depictions of androids and AI, which routinely conflate science and fiction. Laden with ideologies and bringing into play the disposition to anthropomorphise other-than-human entities, worn-out references to robots continuously position them as our servants or beings to be feared as they might rise against us. Relatedly, these technologies are often presented as more advanced, autonomous or humanlike than they are. The issue is that these enduring representations form the backdrop against which robots and AI are being generated and evaluated, shaping both technical aspirations and public acceptance of different automated systems. In effect, these limited interpretations of mere admiration, fear or awe sustain tenacious anthropomorphic tropes that reinforce narrow sociotechnical imaginaries and, crucially, obscure the accurate operations of the industry and accompanying material practices of AI's development and presentation. Put differently, these portrayals of robots and AI perpetuate verbose claims about their imminent replacement, passing for, or surpassing of humans, which distract from the ethical, social, and political implications the expansion of such technologies (many of which are still in their infancy) entail.

DOI: 10.4324/9781003475972-3

Therefore, this chapter explores the creative potential embedded in the robot's incapability rather than lionising big-tech projects as heroic accomplishments, romanticising the autonomous machine or yielding to suspicion and scepticism. Studying the cultural languages enacted in the performance of seemingly faulty automation, I foreground its capacity to probe some of the myths surrounding robots and AI, which can foster alternative understandings of machines. To do so, I focus on the clunky and awkward. I concentrate on the glitchy performances in Rimini Protokoll's *Remote X* (2013–) and *Uncanny Valley* (2018–) and emphasise the generative potential of those slippages. I analyse these deliberate designs of imperfections or unexpected operations to suggest that they can facilitate less fixed or concrete views of machines and thus tamper with anthropomorphism. Challenging airbrushed portrayals, in turn, troubles some of the underlying assumptions and beliefs that drive robotic development and design and can therefore modify HRI and transmute techno-utopian or dystopian fantasies.

Both productions were conceived and directed by Rimini Protokoll's Stefan Kaegi and give the impression of using advanced AI and robotic technologies.[1] *Remote X* is an audio-walk/performance where audience members with headphones are led through the city, guided by a synthetic voice akin to virtual assistants like Siri or Alexa. In *Uncanny Valley* an animatronic double of writer Thomas Melle (who dubs his look-alike) delivers a performance lecture and recounts the biography of computer scientist Alan Turing and Melle's own dealings with bipolar disorder. Importantly, despite their apparent spontaneity, both productions are scripted and offer a "carefully framed and controlled" environment to reflect on encounters with humanlike machines (Sandry 2015, 18). Since audiences are mere spectators and are unable to influence the outcome of either performance, the communication/interaction is one-sided. Nevertheless, both productions contribute to the apparent presentation of robotic abilities more widely. This is similar to other staged encounters in laboratories or theatres, and the latter in particular "can provide fictitious environments to stimulate a suspension of disbelief" (Demers and Horakova 2008, 444), typically applied to the machine's skills and personality regardless of their accuracy. Though this often fuels anthropomorphism, it also provides fertile ground to examine relationships with these technologies and re-establish those relations in the first place.

To an extent, the ambivalence between similarity/otherness, humanlike/machinic, and ability/inability runs through both performances. Such a view also tends to guide recent analyses of social robotics and HRI in/through theatre studies and associated fields (see LePage 2021, Otto 2021, Sone, forthcoming). Previous accounts have examined these productions addressing the metanarratives they depict, the performance of the city, how these reflect on the human condition as well as related questions of affect and embodiment that inform my analysis below (see Martin 2014; Thomaidis 2017; Tait 2021; Otto 2021; Sone, forthcoming). Of particular relevance to this essay is Ulf Otto's formulation of "the android scenario" that consists of the performance of the glitch. For Otto, notwithstanding the initial "spectacle of likeness" (2021, 361) manifested in the android's presentation, which stems from its resemblance to (some) human spectators, the glitch reinstates the status of the human as different from the anthropomorphic machine. In so doing, it dissolves the "'phantasm of replacement': the idea that 'they' might actually replace 'us.'" (362) – and affirms, the somewhat problematic universalist and essentialist notion of the unmarked human.

Despite the shared interest in the glitch, I adopt a slightly different view of AI narratives, to consider not so much the reinforcement of the human status, but to underscore the mismatch between fantasy and actuality. Looking through the attempt for mimetic semblance to humans, I study the glitch to demystify the presentation of independent humanoids. Evaluating acting androids, Otto conceptualises the glitch as "a punctual malfunction [that] discloses the character to be nonhuman and thereby implicitly grants the machine the status of defective and deviant human" (2021, 363). Like Otto, I concur that glitches might unsettle different processes of anthropomorphism. Yet, I approach glitches in more generative terms. As theatrical tactics and dramaturgical devices, I argue that they can facilitate alternative perceptions of machines themselves. More specifically, by agitating anthropomorphism, glitches disrupt the suspension of disbelief and hint at the operative "black box" devised by the makers, exposing the human behind the machine. So, instead of ratifying the machine as a "defective human," I turn to glitches to disclose the android as a seemingly defective *machine*. Therefore, in what follows, after a brief framing of AI narratives and their connection to the industry and presentations of robots and AI, I explore specific glitchy compositions in the aforementioned productions and how they reconfigure rehearsed interpretations of these technologies.

Contemplating Machines

Given the rapid development of AI, algorithms, machine learning, and big data, an expanding body of literature seeks to critique underlying assumptions and reveal the broader societal impacts of these technologies. These framings become significant since, as Meredith Broussard posits, "the way people talk about technology is out of sync with what digital technology actually can do" (2018, 6). While robots, AI, and automation populate various industries and environments, the desire to create and form connections with anthropomorphic figures and humanlike machines is far from new. This is evidenced by a rich legacy of tales and stories ranging from Greek mythology's Pygmalion and the Jewish Golem to Mary Shelley's *Frankenstein* (1818) and Karel Čapek's *R.U.R.* (Rossum's Universal Robots) (1920). Čapek's play not only contributed to the coinage of the term "robot" but also fuelled the association of robots with synthetic/artificial humanlike figures, forced labour, inevitable revolt, and uprising (Sandry 2015, 3; Sone, forthcoming).

With the meteoric rise of their technical development, robots and AI continue to occupy public discourses and news cycles, which are informed by and mirrored in the proliferation of their appearances in literature, films, artworks, and growing presence in theatre performances. These fictional representations form part of what has been termed "AI narratives" (Cave et al. 2020) referring to the various ways these technologies are and have been imagined, portrayed, discussed, and perceived. Regardless of accuracy, these technologies are habitually cast as human-others – most notably as servants, lovers, or possible overlords – leading to the formation of certain cultural imaginaries. Through well-known characters like HAL9000, C-3PO, WALL-E, Maria, Samantha, and Ava, which are still fantastic Hollywood fables, frequent allusions to science fiction tropes ossify specific perceptions of machines and their abilities, which distort the expectations of users and designers alike.

Notwithstanding their seemingly autonomous façade, the reality is, as Broussard notes, that "[f]or every AI system that exists today, there is a logical explanation for how it works" (2018, 33). Yet, the presentation of – especially humanoid – robots as lifelike or intelligent relies on "a series of erasures that work to obscure the material practices and extended relations on which the agencies not only of robots, but also humans depend" (Suchman 2019, 56). For example, Shanyang Zhao's early assertion, that "A revolution of humanoid social robots is quietly taking place in our society: autonomous, interactive and humanlike entities of various sizes and shapes are leaving research laboratories in large numbers, making their way into the world of our everyday lives" (2006, 402) is prototypical. This statement, both hyperbolic and unassuming, suggests the machine's independence, omitting the human programmers and designers responsible for this so-called revolutionary shift. Thus, instead of repeating familiar celebrations of robotic abilities or glorifying technological advancements, this analysis turns to performance practices to contribute to the growing scholarship that seeks to look behind the curtain. It examines the legacy and implications of AI narratives: how they structure and influence contemporary discourse around robots/AI, while simultaneously misrepresenting industry practices and operations (see Suchman 2007, 2019; Heffernan 2019; Cave et al. 2020; Coeckelbergh 2022). Similar to Otto's previously mentioned thesis, these accounts identify how robots often sustain a universalising and human-exceptionalist progress narrative that reiterates "a commitment to the figure of the human subject characterized [*sic*] by bounded individuality and to the reproduction of an order of hierarchical humanity deeply rooted in imperial/colonial histories" (Suchman 2019, 36). By inspecting the supposed "neutrality" and "universality" of robots, they become "disclosing agents" (Suchman 2007, 241), revealing how they reflect broader cultural values and hegemonic narratives that replicate existing prejudices, inequalities, biases, and exclusions.

Furthermore, these figurations are regularly dominated by transhumanism – the belief in the radical enhancement of the human condition through technology; the inevitability of strong or "general" AI; the emergence of an autonomous machine superintelligence that will surpass human intelligence; the technological singularity; and the merging of man and machine to achieve immortality (Heffernan 2019, 3). Disturbingly, as Teresa Heffernan argues, this "religious-like faith in technology" leads those overseeing the technology and tasked with preventing it "from going rogue" to become deeply committed to its potential for profoundly improving humanity. This results in a privileging of algorithmic logic and big data, and the tendentious positioning of humans as "autonomous machinelike managers and masters of the cosmos rather than animals that are dependent on the planet" (3). Far from being a modest technological fascination, this drives the robotics industry to pursue very particular and hubristic trajectories.

Consequently, led by media hype and guided by corporate, economic, or governmental interests, numerous discourses revolving robots and AI heavily invest in the promises or dangers of these "magical" technologies. However, the focus on whether machines will one day go rogue, replace humans, and take over the world – and whether these forecasts are exciting or alarming – distracts from current controversial operations of big data companies and the problematic infrastructures that propel robotic research and design. These issues might include data collection sold to third parties, job displacement, the promotion of problematic assumptions about marginalised groups through generated material, exploitative labour and production practices, and environmental impacts to facilitate computational power, among others. As Heffernan details, "the very algorithms that are driving the industry are necessarily informed by the biases of coders and the prejudices of the human-created material they mine, and they replicate all the toxic inequities that have infected human culture" (2019, 5–6).

Critically, due to the "black box effect" – where data inputs lead to outputs without transparency regarding their connection – it becomes nearly impossible to challenge or question the reasons behind a machine's actions. This can promote claims of machines running amok but, together with their presentation as independent entities, "deflects responsibility and complicates industry liability" (5).

Indeed, each robot (or machine learning software) has a designer behind it. Yet, to enable more "authentic" interactions with machines and facilitate the social robot's full effect and seemingly smooth operation, programmers are often concealed alongside the back-end mechanisms. This fosters "the illusion of intelligence" or "leads users to believe in the sophistication of the system in areas where the user will not encounter its failings" (Duffy 2003, 178–179), which, paradoxically, undermines those encounters, rendering them insincere. Therefore, when it comes to HRI, there is usually an inherent "uneven distribution of knowledge about the real state of affairs" (Coeckelbergh 2022, 2051). While the developer "knows the reality," this repetitive gesture towards the machine's humanity or seamless functionality leads users at best to "temporarily suspend disbelief and embrace the illusion" and at worst to be "misled to believe that the robot is a person" (2051–2). Challengingly, this increases user expectations and has a notable influence on the robot's reception. For Broussard, "understanding the computational logic can demystify AI" (2018, 33), which offers a strategy to destabilise the dominance of the rehearsed portrayals as it uncovers concrete processes and reasonings. Intervening in the systematic operation of the machine points to the potential of glitches, which, I argue, can equally unearth the inaccuracies or exaggerations embedded in formulations of AI. I'll now turn to *Remote X* and then discuss *Uncanny Valley*.

Remote X, Glitching Synthetic Voices

In *Remote X*, 50 audience members set out to explore a city guided by a synthetic voice. Since its premiere in Berlin in 2013, the production has toured over 50 cities from Lisbon and Riga to Tehran, Taipei, and LA, revealing the mechanical operations of daily urban life. Adapting to each performance location, in *Remote Berlin*, the digital voice of Rachel directs listeners through streets, parks, cemeteries, churches, and hospitals. As they venture into urban locales, audience members are encouraged to observe this city from the perspective of artificial intelligence as the production questions the implementation of AI and big data as well as our own predictability and mechanical tendencies. To cite Rachel: "Every traffic light is a small exercise in automated dictatorship – a game between the human and the machine" (Rimini Protokoll 2015 [2013]). A disembodied synthetic voice, Rachel has no physical figure. "Why don't you close your eyes and try to picture me?" she prompts; "How do you imagine my eyes looking at you? Do you find my voice strange? I have no lips; I have no mouth" (Rimini Protokoll 2015). Rachel's performance received praise for its believability, as in Carol Martin's observation that "The eerily realistic quality of Rachel's voice created an illusion of real presence" (2014, 13). Yet, the synthetic voice does not conceal its artificiality – "I'm software, an idea" (Rimini Protokoll 2015). Rachel combines snarky commentaries with contemplative and friendly invitations, and as the audience, we must trust this artificial entity to guide us safely. "I don't have a body," she reminds us, "but I know what a car can do to yours" (Rimini Protokoll 2015).

Synthetic voices have existed for quite some time. More recently associated with GPS navigators, virtual assistants like Siri and Alexa, call centres or video games, early voices were based on "a real voice actor, whose likeness (with consent) has been preserved using AI" (Hao 2021, n.p.). Initially, the speech was generated by gluing phonemes, syllables, and sounds, which achieved somewhat clunky affectation that has become concomitant with robotic voices. Karen Hao explains that making these voices "sound any more natural was a laborious manual task" (2021, n.p.). Deep learning breakthroughs now allow developers to input large amounts of auditory material, enabling algorithms to learn patterns independently, and replicating some of the nuances of human speech, including pauses, breathing, and changes in style or intonation. While extended speech can reveal its artificial nature, some short audio clips "have become indistinguishable from humans" (Hao 2021, n.p.). Akin to deepfakes, these tools raise ethical issues about authority and trickery, though claims of indistinguishability often carry the hyperbolic and alarmist rhetoric about human replacement.

Rachel's monotonous voice is far from representing state-of-the-art software. However, despite occasional self-deprecating comments, it holds power over the experience, functioning as the seemingly trustworthy guide, who carefully leads the audience. "You plan your way," Rachel notes, "but I am directing your steps" (Rimini Protokoll 2015). Thus, the voice becomes performative: not merely conveying semantic meanings but also constructing the experience scenographically and dramaturgically and casting itself and the audience in their respective positions. In this way, it performs various roles beyond simply demonstrating technical abilities. This chimes with Louise LePage's assertion that "robots are inherently performative; their identities derive entirely from their performances" (2021, 1430), i.e., the sum of their actions. While lacking inherent personality or identity, they can be engineered to appear as if possessing such qualities (1429). Konstantinos Thomaidis recounts his experience of *Remote X*, illustrating how the voice sounds "monotonously pleasant and alarmingly reassuring," and as the journey unfolds, it becomes "personable, poetic, instructive, authoritarian, malleable, untrustworthy and uplifting" (2017, 7). Indeed, a key aspect that renders the "human voice so human is its inconsistency, expressiveness, and ability to deliver the same lines in completely different styles, depending on the context" (Hao 2021, n.p.). In *Remote X*, while the repetitive inflexion of the synthetic voice might serve as a strategy to foreground its artificiality, it seems the tendency to anthropomorphise the machine trumps this perception. Importantly, part of what fosters this rapport or affinity towards the artificial voice is the acclimatisation to Rachel, "where all the parts coalesce to produce a socially identifiable figure demonstrating predictable behaviour," which results in an "aesthetically coherent" representation of the character (LePage 2021, 1427).

But then things change. Towards the end of the tour, Rachel announces, "Let me surprise you. I am going to transform now. I am going to turn into Peter. That is me too, even when Peter sometimes claims otherwise. I will start the metamorphosis now" (Rimini Protokoll, 2015). From then, the mechanical performer repeats the phrase "I am Rachel," glitching, and gradually mutating into "I am Peter," completing the transformation with "My name is Peter, don't worry everything has been saved. Nothing is lost, everything is normal. I am sure that we will work together just as well as you and Rachel did" (Rimini Protokoll, 2015). This move occurs – somewhat ironically – in a church, tapping into notions of morality, judgement, and decision-making humans regularly grapple with but mark clear limitations for the machine. Nevertheless, the audience must now engage with Peter, forced to form a connection and trust a new artificial agent. This moment of transformation "opens a gap between what is said, who is saying it, how things are told and crucially how things are understood or could be understood" (Kendrick and Shyldkrot 2021, 149). It oscillates between different bodies and meanings. Hence, bringing about "a less fixed listening experience" (151) where the seemingly coherent "character" becomes – even momentarily – incoherent.

There is an inherent paradox in designing glitches since the software ends up properly executing its intended breakdown, so it is arguably not a glitch in its familiar interpretation. However, this planned malfunction expands "the understating of glitches from accidental errors in the system, to intentional interventions," in this case, in the aural sphere (Kendrick and Shyldkrot 2021, 148). Rather than a bothersome or unwelcome disruption, as theatrical tactics highlight uncertainty, these deliberate designs of imperfection have the potential to unsettle or problematise expected utterances or perceptions. In *Remote X*, as Rachel/Peter voluntarily and surprisingly breaks down this glitch destabilises different processes of anthropomorphism by performing the machine's limitation, even more than Rachel's emphasis on its artificial nature.

At one level, the shift from Rachel to Peter creates an opportunity for gender fluidity that begins to dismantle expected binaries and norms manifested in robot design, particularly in the gendering of virtual personal assistants. This relates to wider processes where social robotics produce certain interactions with machines by confirming problematically assumed behaviours, roles, and expectations. For example, gendering a household/service robot female reinforces a culture where women are supposedly expected to take on such roles (Coeckelbergh 2022, 2055). This glitching of the female voice disrupts the coherence of the character, its assumed features and functions, and consequently challenges the ways it is anticipated to perform.

More broadly, though the voice has a name and conveys an assumed character, it is not attached to a physical body. *Remote X* continuously points out this nonexistence. As Rachel clarifies: "I didn't create you, you created me. You look at an image and imagine me. But that image is not me. I leave no trace" (Rimini Protokoll, 2015, n.p.). This links Rachel/Peter with the concept of the acousmêtre – a sound one hears without seeing its source. Initially prominent in cinema, acousmatic/disembodied voices have become common in a ubiquitously digitised culture. Within it, the talking computer/machine presents a tricky form of the acousmêtre. As Liz Faber argues, AI can be seen as embodying the computer from which the voice seems to emanate. However, "the computer's voice is a prime example of the disordered voice, separated from its origin (a real human voice actor) and projected onto the image of a computer object, though maintaining an indexical link to the actor's gendered body" (2020, 15). In other words, there are attempts to return to an emitting body (of the performer or "embodied" machine), while simultaneously "effacing the workings of the apparatus" (11). This aligns with contemporary robot presentations, which, as noted, conceal the human behind the machine. Correspondingly, tensions between physical embodiment and machinic voice, enabled by technological reproduction, intertwine with fantasies about disembodied voices and their supposed powers (Thomaidis 2017, 61) solidifying AI narratives and imaginaries. The denial of synchronisation of voice-body becomes more acute in *Remote X*, where the disembodied voice loses sync further. Albeit software-generated, growing accustomed to Rachel, listeners may "have started invoking a vocalic body for her" (66), which the metamorphosis into Peter shutters. The robotic voice remains at large, unable to land or return to a secure figure. As Rachel glitches into Peter, "[v]oices and bodies are stretched and reformed, extended and fragmented," the familiar signal disintegrates and the performance of a stable identity crumbles (Kendrick and Shyldkrot 2021, 153). The inability to return to a fixed entity challenges anthropomorphism, underscoring the artificial and staged qualities of the performance that undermines claims of the machine's autonomy or independence and thus demystifies AI.

Uncanny Valley, Performing Breaking Androids

Sitting in an armchair with his legs crossed, a mechanical replica of German author Thomas Melle narrates the life of computer scientist Alan Turing and Melle's personal experiences with bipolar disorder. As he considers some of the mechanics of mental illness, Melle contemplates his possible replacement by this ostensibly sophisticated and autonomous ventriloquist's puppet, speaking in his voice. *Uncanny Valley* is a direct reference to Masahiro Mori's 1970s theory of uncanny valley – where human's affinity to artificial objects increases the more humanlike they become, until the moment when the object becomes too humanlike which makes it uncanny. Wrestling with the appearance and acceptance of androids, the performance offers a valuable example to my study of HRI in relation to common imaginaries and speculations of replacement by machines.

The emphasis on appearance is crucial here. Otto elucidates that technically the performance has little to do with AI, and the mechatronic-double of Melle is simply a dubbed animatronic device – "the same technology that Steven Spielberg used to animate his dinosaurs, devoid of any machine-learning functionality" (2021, 362). However, in the same vein that the synthetic voices in *Remote X* were representations and not truly interactive software operated by generative AI, so is the performance of the Melle-machine. Its presentation – akin to recent humanoid robots like Sophia, Pepper, or Ameca – is suggestive of greater capacity than the machine's exact abilities. These demonstrations give the impression of an enhanced ability for independent thought, conversation, responsiveness, intentionality, emotions, or intelligence, contributing to the assumption that "AI has advanced further than it has in reality" (Złotowski et al., 2015. 352). Though stage robots differ from social robots intended for real-world application, as the former act as placeholders for representation, the fact that "[b]efore it says or does anything, audiences invest the humanoid robot with the promise, if not the fact, of intelligent autonomy, character, and life" (LePage 2021, 1430) encourages inaccurate or distorted perceptions about their capabilities. Once again, this highlights the potential of glitches to unsettle anthropomorphism and demystify the presentation of independent humanoids. The glitch becomes a bug in the unity between reality and mimesis. Disrupting the suspension of disbelief, these dramaturgical tactics and deliberate designs of imperfection unearth the pre-programmed nature of the performance, pointing to the humans behind the machine and thus reconfiguring perceptions of autonomy.

In incorporating a double, *Uncanny Valley* reflects on HRI through physical embodiment. The Melle-robot's "realistic" or "lifelike" appearance includes details like moles, hair, or stubble and the ability to emulate different facial gestures. In this, the production considers the disposition to anthropomorphise robots and machines. Brian Duffy defines anthropomorphism, as the tendency of "ascribing human-like qualities to a system based on one's interpretation of its actions" (2003, 181). In robotics, this adaptive trait has become a means of enabling interactions and a tool for integrating robots in various scenarios (Złotowski et al. 2015, 352). This has a bearing on the designer's approach to their object's appearance as well as other cues and affordances exhibited during the encounter that shape the machine's actions and conjure its behaviour. Notably, the key is not whether a system actually possesses these abilities, but whether it displays traits that encourage people to perceive or interpret it as having them – be it autonomous, intelligent, friendly, and so on (Duffy 2003, 179). Put differently, as the machine gives the impression of holding particular qualities, it relies on the performance of properties commonly associated with those characteristics (e.g. being alive or intelligent), suggesting that these are not inherent or natural abilities but a performance of socially and culturally coded characteristics. On the surface, *Uncanny Valley* complies with familiar tropes of robots and humans becoming interchangeable and the former's growing capacities. Yet, compositionally, the performance continues to expose the mechanisms, hindering the attribution of set or strictly human characteristics to the mechanical double. These compositions unfix anthropomorphic displays and challenge the expectation that the robot is or will be sentient and self-directed.

While Turing poetically proposed that designing humanoid machines "would have something like the unpleasant quality of artificial flowers" (2004 [1951], 486), common industry practices seem guided by the principle that

> the more humanlike a robot is in its behaviour, in its appearance, and in the manner with which it interacts with us, the more ready we will be to accept it as an entity with which we are willing or even happy to engage.
>
> *(Levy 2007, 13–14)*

Further, David Levy contends that "the eventual acceptance of robots as sentient beings, worthy of our friendship, our love, and our respect will be greatly facilitated by the physical design and construction of robots whose appearance matches our notions of friendliness" (159). This view, suffused with human-exceptionalism, risks the development of false expectations and limits engagements with diverse robotic forms, since humanoid robots, by their design, compel specific interactions aligned with their anthropomorphic traits. Eleanor Sandry explains that this perspective is "strongly reinforced by the widely accepted understanding that communication is based on what communicators have in common, with its success being related to increasing that commonality further" (2015, 13; Maguire-Rosier et al. 2024, 89). In *Uncanny Valley*, Melle asserts this outlook, noting that "Machines can be friends as long as they look very similar to us. The eyes of such machines must be able to look at us. Their voices must come from their lips" (Rimini Protokoll 2019). Importantly, by valorising the significance of commonality for successful communication, humanoid robots reinforce certain Western understandings of "universal" expressions that do not account for cultural differences, pursuing predisposed expectations of what constitutes both desired interactions and natural/neutral human (usually white, male, and able-bodied). Notably, although Melle (and his double) are male and white-presenting (see Otto 2021), the production can be seen as challenging the fixity of categories and the attribution of these human characteristics to mechanical objects.

If in *Remote X* the artificial voice had no body to return to, the android in *Uncanny Valley* is dubbed by the voice of the "real" Melle, establishing, as Yuji Sone proposes, "modalities of ventriloquism" (forthcoming). Though this might relegate the machine to a dummy/puppet status, it initially supports the anthropomorphic presentation of the Melle-robot that is "constructed when a voice matches an image of a person's lips moving, thereby creating the sense that a person is speaking within the diegetic space" (Faber 2020, 14). Cue the glitch. Though at first, Melle's voice seems to "come from their lips," as the performance unfolds, the mouth freezes and mouth movements and voice go out of sync. Self-referentially, Melle reveals that his double was equipped with throat-clearing sounds and slips of the tongue to make it sound more "natural." However, much like Rachel's glitch-transformation, Melle-robot incessantly fails, upsetting cohesion and generating less fixed or assured perceptions of the machine (Kendrick and Shyldkrot 2021, 151). This is complicated further when human-Melle appears on video and animates his mechanical doppelganger. Practically, in this relatively "hierarchical relationship," the performance of each reveals the machine's "qualitative differences from the human" (Sone, forthcoming). Yet, like Narcissus transfixed by his reflection, it seems that in using the same (human) voice for both representations, *Uncanny Valley* forms a more complex relationship between copy and original: the android-double and "real" Melle, the removed 2D presence projected on screen, and the live representation of the author: "Now I am here and not here," he notes (Rimini Protokoll 2019). Held together by the voice, both figures index a real body that is not there, uncovering the artifice and blurring distinct categories.

Presentations of robots often encourage users to overlook the robot's limitations or choppy movements. LePage argues that when robot performers are cast as humanlike characters, they can "compensate for flaws in the physical performance of human likeness" (2021, 1432). However, intentionally or not, Melle-robot does not solely attempt for smooth motion, but glitches and moves awkwardly throughout. The legs jerk, hands move clunkily, and at some point, the facial features slip, with eyes and mouth dropping and suspended, seemingly losing control. While anthropomorphic depictions strive for optimality in robot motion, Melle's physical performance routinely stunts the illusion. Similar to the approach for robots' appearance and physical features, "motion design in social robotics has remained fixated on anthropomorphic or human imitation-based concepts" (Maguire-Rosier et al. 2024, 89). Utilitarian characteristics of speed and accuracy notwithstanding, the performativity of robot motion is an equally vital factor during HRI (Maguire-Rosier et al. 2024; Abe, 2022). This becomes significant when screen-Melle instructs the android to twist their ankle 180 degrees. The foot then rotates inward at an impossible angle, accompanied by the theatrical sound of cracking bones. As Melle ponders on his looming replacement, the foot continues to shake and swivel. Albeit a planned malfunction, this glitch marks the machine's inability to "properly" perform *as a human*, while simultaneously signifying the possibility that machines can also act in visibly non-human ways (Sone, forthcoming). Despite an ostensibly impressive first impression, these deliberate faults do not simply portray a defective human; they successfully perform their breakdown, paradoxically representing a seemingly defective machine. Highlighting their mechanical, pre-programmed, and performative characteristics, these glitches challenge anthropomorphic perceptions, constantly reminding the audience they are engaging with a machine that might not be as capable or autonomous as often thought.

Conclusion

This chapter explored how the glitchy performance of robots and AI can trouble HRI. Rather than seeing these breakdowns as unwanted, I highlighted the potential of these performative interventions to destabilise expectations of seemingly autonomous machines, revealing new possibilities for interaction beyond dominant sociotechnical imaginaries and thus highlighting the significance of theatre and performance practices to the study of HRI.

Despite recent advancements, machines and AI systems are still fundamentally dependent on human design, programming, and oversight. While robots and AI can execute tasks with impressive speed and accuracy, their capabilities are limited to what they are explicitly programmed or trained to do. Behind their deceptive autonomy lies a complex network of algorithms, data inputs, and predefined rules set by humans. Anthropomorphising machines may be an inevitable part of human meaning-making (Coeckelbergh 2022, 2058). However, given current industry practices that replicate biases and overstate technical capabilities, it is crucial to interrupt habitual perceptions and rethink our relationships with machines.

Both *Remote X* and *Uncanny Valley* exemplify this by performing several glitches, dislocations, and disruptions. Through these, Rimini Protokoll tears and defies synchronicity, disjointing the expected unity of voice and mouth, body and expression, utility and movement, as well as the conventional coupling of robots with familiar tropes and rehearsed narratives. Specifically, Rachel/Peter and Melle's robot become performative characters dislodged from longstanding formulations of machines as capable slaves or threatening masters. By tampering with anthropomorphism, the company opens space to question who we are interacting with and how, revealing the humans behind the machines and underscoring the importance of considering less common, messy, or modest portrayals of machines. This approach is pivotal to developing alternative perceptions of these technologies. Embracing these complexities and imperfections supports fundamental reconsiderations of how both robots and humans are conceived, leading to more nuanced, ethical, and meaningful interactions and technological projects.

Note

1 Robots and AI refer to distinct technological projects with different applications, functions, and embodiments. However, the promises, fears, and narratives involving both are often conflated. Since each of my case studies explores AI and robots respectively, and the critiques raised apply to both, I use the terms interchangeably.

References

Abe, Naoko. 2022. 'Beyond Anthropomorphising Robot Motion and towards Robot-Specific Motion: Consideration of the Potential of Artist—Dancers in Research on Robotic Motion'. *Artificial Life and Robotics* 27 (4): 777–85.

Broussard, Meredith. 2018. *Artificial Unintelligence: How Computers Misunderstand the World*. Cambridge, Massachusetts: MIT Press.

Cave, Stephen, Kanta Dihal, and Sarah Dillon, eds. 2020. *AI Narratives: A History of Imaginative Thinking about Intelligent Machines*. Oxford: OUP.

Coeckelbergh, Mark. 2022. 'Three Responses to Anthropomorphism in Social Robotics: Towards a Critical, Relational, and Hermeneutic Approach'. *International Journal of Social Robotics* 14 (10): 2049–61.

Demers, Louis-Philippe, and Jana Horakova. 2008. 'Anthropocentrism and the Staging of Robots'. In *Transdisciplinary Digital Art. Sound, Vision and the New Screen*, edited by Randy Adams, Steve Gibson, and Stefan Müller Arisona, 434–50. Berlin: Springer.

Duffy, Brian R. 2003. 'Anthropomorphism and the Social Robot'. *Robotics and Autonomous Systems* 42 (3–4): 177–90.

Faber, Liz W. 2020. *The Computer's Voice: From Star Trek to Siri*. Minneapolis: University of Minnesota Press.

Hao, Karen. 2021. 'AI Voice Actors Sound More Human than Ever—and They're Ready to Hire'. *MIT Technology Review*. 9 July 2021. Accessed 22 May 2024. https://www.technologyreview.com/2021/07/09/1028140/ai-voice-actors-sound-human/

Heffernan, Teresa, ed. 2019. *Cyborg Futures: Cross-Disciplinary Perspectives on Artificial Intelligence and Robotics*. Cham, Switzerland: Palgrave Macmillan.

Kendrick, Lynne, and Yaron Shyldkrot. 2021. 'Voicing Identity Theatre Sound and Precarious Subjectivities'. In *Avatars, Activism and Postdigital Performance: Precarious Intermedial Identities*, edited by Liam Jarvis and Karen Savage, 137–60. London: Bloomsbury.

LePage, Louise. 2021. 'The Importance of Realism, Character, and Genre: How Theatre Can Support the Creation of Likeable Sociable Robots'. *International Journal of Social Robotics* 13 (6): 1427–41.

Levy, David. 2007. *Love and Sex with Robots: The Evolution of Human-Robot Relationships*. London: Gerald Duckworth & Co Ltd.

Maguire-Rosier, Kate, Naoko Abe, and Fiona Andreallo. 2024. 'What Other Movement Is There?: Rethinking Human-Robot Interaction through the Lens of Dance Performance'. *TDR* 68 (1): 87–103.

Martin, Carol. 2014. 'Performing the City'. *TDR: The Drama Review* 58 (3): 10–17.

Otto, Ulf. 2021. 'Performing the Glitch: AI Animatronics, Android Scenarios, and the Human Bias'. *Theatre Journal* 73 (3): 359–72.

Rimini Protokoll. 2015. 'Remote Berlin - Documentation (Stefan Kaegi and Jörg Karrenbauer)'. https://vimeo.com/116552002

Rimini Protokoll. 2019. 'Uncanny Valley - Documentation (Stefan Kaegi)'. https://vimeo.com/339074946

Sandry, Eleanor. 2015. *Robots and Communication*. New York: Palgrave Macmillan.

Sone, Yuji. [Forthcoming]. 'Android Doubles in Performance: Contrasting Conceptualizations of the "Self"'. In *Revealing Posthuman Encounters in Performance*, edited by Sarah Lucie and Stefano Boselli. London: Routledge.

Suchman, Lucy. 2007. *Human-Machine Reconfigurations: Plans and Situated Actions*. Cambridge: Cambridge University Press.

Suchman, Lucy. 2019. 'Demystifying the Intelligent Machine'. In *Cyborg Futures: Cross-Disciplinary Perspectives on Artificial Intelligence and Robotics*, edited by Teresa Heffernan, 35–61. Cham, Switzerland: Palgrave Macmillan.

Tait, Peta. 2021. *Forms of Emotion: Human to Nonhuman in Drama, Theatre and Contemporary Performance*. London: Routledge.

Thomaidis, Konstantinos. 2017. *Theatre and Voice*. London: Palgrave Macmillan.

Turing, Alan. 2004. *The Essential Turing: Seminal Writings in Computing, Logic, Philosophy, Artificial Intelligence, and Artificial Life, plus the Secrets of Enigma*. Edited by B. Jack, Oxford: OUP.

Zhao, Shanyang. 2006. 'Humanoid Social Robots as a Medium of Communication'. *New Media & Society* 8 (3): 401–19.

Złotowski, Jakub, Diane Proudfoot, Kumar Yogeeswaran, and Christoph Bartneck. 2015. 'Anthropomorphism: Opportunities and Challenges in Human–Robot Interaction'. *International Journal of Social Robotics* 7 (3): 347–60.

3

MEDUSAI

A Multi-modal Large-Scale Robotic Musician

Amit Rogel, Jiahe Qian, Ripken Walker, Nicollete Cash, Emily Liu, Hope Phan, Hannah Schlisky, Tristan Al-Haddad and Gil Weinberg

3.1 Motivation

Recent advancements in generative AI have led to significant societal shifts in artistic expression among many other aspects of human life. Such a significant societal shift is often accompanied by and sometimes a generator of myths – stories that debate, reveal, and pass on culture's aesthetics, values, and tensions. One of the most prevalent myths in Western culture has been the Ancient Greek myth of Medusa, who was born as a mortal maiden known for her beauty and good nature. Medusa was raped and abused by Poseidon and cursed by Athena to become a snake-haired monster who turns anyone who looks at her into stone. She continued to bestow death, even after Perseus, son of Zeus, beheaded her. The medusai project builds on the parallels of this story to the current effect of AI on society. If the virtuous, well-intentioned motivation for AI refers to the young Medusa, who in our story might play the roles of Poseidon, Athena, and Perseus? The project also relates to other mythological snake references prevalent in Western culture. One of the most well-known is perhaps the story of the snake in the Garden of Eden, who tempted Eve, the archetypal mother of humanity, to eat from the forbidden fruit. Here too, the snake is both seductive to humans, embodying terrible dangers of explosion and death.

Based on the questions and artistic statement posed above, we defined the following creative goals for the project: 1. Design a sculpture that projects uncanny and unnerving aesthetics; 2. Develop interactions that create

DOI: 10.4324/9781003475972-4

positive creative and artistic engagements; 3. Build a large-scale Greek-inspired sculpture that supports interactions with multiple users; 4. Combine modern and ancient design elements that bring the Ancient Greek aesthetics, one of the bedrocks of Western civilisation, to our current Western-driven technological landscape.

To address these goals, the kinetic sculpture incorporates seven robotic arms on top of a human-size head structure. Similar to Medusa's snake hair, the robotic arms inspire awe while sparking fear at the same time. The project also utilises embedded speakers and lights, room-size projected animation, a surround audio system, and ceiling-mounted lights to create a large-scale immersive experience.

Three forms of musical outputs have been incorporated into the structure: 1. robotic arms plucking strings, 2. robotic arms playing the drums, and 3. spatial speakers which complement the acoustic sounds with surround electronic music. We also utilised lights and robotic gestures as additional forms of output. For inputs, medusai utilises a camera-based vision system that tracks the number of people around it, including their location and gestures. Acoustic pickups are embedded around the structure for novice and professional musicians to engage in music-making through plucking the strings and drumming on the sculpture surface.

3.2 Related Works

The field of "robotic musicianship" is often divided into two research areas – "musical mechatronics," which is the study and construction of physical systems that generate sound through mechanical means; and "machine musicianship," which focuses on developing algorithms and cognitive models representing various aspects of music perception, composition, performance, and theory (Weinberg et al., 2020; Bretan and Weinberg, 2016):

Robotic musicians can be designed to play traditional instruments such as piano (Jaju et al., 2016; Zhang et al., 2009), flute (Kato et al., 2018; Solis et al., 2007), guitar (Yang et al., 2023; Davis, 2007), marimba (Hoffman and Weinberg, 2010; Zahray et al., 2020), or drums (Liu et al., 2018; Weinberg et al., 2006). They can also be designed to play "New Interfaces for Musical Expression (NIME)" or have sound creation implemented into the robot itself in the form of "Mechatronic Musical Instruments (MMI)" (Gurevich, 2014). For example, Placencia and Carnegie created a modular string-playing device titled Azure Talos – a tabletop structure containing six strings that are excited by a revolving picking mechanism (Yepez Placencia et al., 2020; Carnegie et al., 2020). While Azure Talos can generate a variety of plucking sounds, its mechanical gestures are limited to repetitive small movements that cannot support gestural musical expression. Vindriis' StrumBot, on the other hand, uses a robotic arm to strum a pitch-shifting string, which can project some expression in its ancillary gestures (Vindriis and Carnergie, 2016). StrumBot, however, is not programmed to respond to human input and does not provide multiple sound generation modalities.

Unlike medusai, previous robotic musicianship projects are often limited in size, focus on a limited sound excitation method, and cannot be experienced interactively by a group. Only a few efforts have been made to integrate mechanical actuators into large-scale musical installations (Flø and Wilmers, 2015; Singer et al., 2005). While these efforts feature some elements of the interactions and modalities explored in medusai, they do not integrate these modalities into a central embodied robotic musician that utilises gestures, music (both electronic and acoustic), light, and projection as featured in this project.

3.3 Sculpture Design

Before designing and constructing the Greek-inspired version of medusai, we experimented with a variety of human-robot interaction schemes using robotic arms from uFactory, a Chinese robot manufacturer, installing them on an 8 ft geodesic dome. The dome was made of steel rods forming the frame, with open faces that were used to mount robotic arms, strings, and percussion surfaces. A camera was embedded at the front of the structure to detect human gestures within a 100-degree field of view. The strings were mounted onto the frame and amplified by custom-built planar speakers. When a robotic arm plucked a string, the steel frame resonated and amplified the sound.

The development of the geodesic dome prototype led to three main goals for the current Greek-inspired design: 1. Design a larger-scale and more accurate computer vision system to capture humans moving around the sculpture; 2. Add audio as an input modality for the interaction; 3. Embed lighting to enhance the experience and guide users around the sculpture. With these objectives in mind, we added the following features to the current design:

1 **Vision:** A wide-angle camera was mounted on the ceiling in front of medusai to capture a wider area of human activity around the sculpture. A YOLO vision system (Huang et al., 2018) which allows for tracking multiple people was implemented to track human activity around the sculpture.
2 **Sound:** Electromagnetic pickups were installed in the structure in order to capture physical impact such as in drumming or plucking. Multiple

speakers were installed in and around the structure to project electronic sound in response to human activity.

3 **Lights:** Strips of multi-coloured LED lights were embedded in the structure in addition to a room-scale projected lighting system designed to respond to detected human activity.

Steel was used for the final Greek-inspired design of the sculpture due to its acoustic properties that allow sound to bounce and echo when chambered. Additionally, utilising steel simplified the sculpture-faceted design, making it easier to mount the robotic arms, instruments, and electronics. Aesthetically, medusai is modelled after the asymmetrical features of Medusa's head, particularly around the eyes and nose areas. To maintain the theme of asymmetry, the following rules were applied to the faceted design: no triangle could have a vertex along the edge of another triangle, no triangles could be equilateral, and no triangles within a "section" could be the same size. Above the hairline, non-uniform extensions were added in the form of steel extensions to resemble low polygon snakes. These steel extensions provided a visual transition between medusai's facial structures and the robotic arms. The extensions were shaped as "snake nest," each consisting of two steel chambers facing each other, with a robotic arm installed in between, and a string mounted between them. Design considerations for the nests focused on robot arm placement and string length. Each string had to have a minimum length of 250 mm to allow a robotic arm to fit through and properly pluck the string. The strings were limited to a maximum length of 600 mm, as this is the maximum length to properly tension our string to play G3 (196 Hz). The top left of Figure 3.1 shows an example of a singular nest. The full combination of seven different nests is shown on the top right. For string plucking, 3D-printed tips were designed to resemble snake heads, with guitar picks embedded at the end to look like tongues, as shown at the bottom of Figure 3.1.

For acoustic amplification, string tuners were mounted at the ends of the steel snakes in each nest. Metal guitar strings were tuned to different frequencies for each of the seven nests. The hollow insides of the steel snakes were partitioned to create volumes that acted as internal resonators to amplify the string sounds. We calculated these volumes using the Helmholtz resonance equation for the natural frequency of air vibration, assuming the neck had a small opening. Each snake nest chamber also contained a speaker.

For electronic amplification, pickups were mounted on the surface of the sculpture. Sounds from hitting around the pickups were processed through effects and played back via the spatialised embedded speakers. Drumming across the structure created a wide variety of timbres, as each

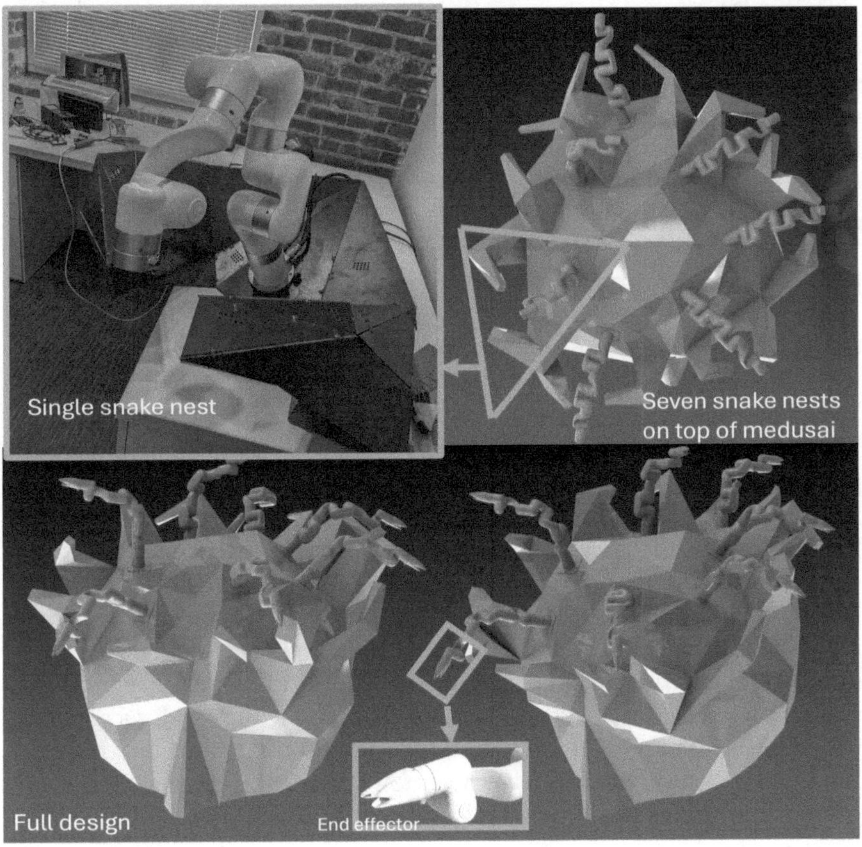

FIGURE 3.1 Medusai's design models: single-snake nest prototype (top right), bird's-eye view of snake nests (top left), end-effector design (bottom centre), and final design renders (bottom left and right). (Credit: Amit Rogel).

facet has a different size and side length, resulting in unique resonant patterns and fundamental frequencies.

3.4 Human-Robot Interactions

The sculpture has two modalities for input – vision and audio – and three forms of output – lights, sound, and robotic gestures. The sculpture functions in two states, one is designed for interactive installations for the general public, and the other for pre-rehearsed performance with professional musicians and dancers. In "installation mode" medusai interacts with any number of human participants that have no prior experience with the sculpture. Here, we chose to focus on highlighting the robotic arms snaking gestures. During installation mode, the robotic arms are moving in a slithering manner to provide engaging visuals for the audience. The autonomous snaking gesture is occasionally interrupted when the robotic arm stochastically plucks a string. On some occasions, a robotic arm will switch to follow a participant who walks near a snake nest. Music for this mode is primarily projected through embedded speakers playing prerecorded tracks. "Performance Mode," on the other hand, is designed for musicians and dancers to practice and publicly perform with the sculpture. In this mode, we focus on medusai's ability to play melodies and rhythmic patterns, which requires the robotic arms to stay close to the strings to play more frequently. While the gestural interaction in this mode is limited, the acoustic music can be played more fluently.

To capture humans' activities around medusai, a top camera has been installed to track movement around the front half of the structure. This is inspired by the mythological Medusa, who could only see what was in front of her. In this area, we track the positions of participants' limbs to determine the number of people in any area, as well as each person's pose. We use Google's YOLO model to get a participant's x,y coordinates around medusai (Huang et al., 2018). Based on the changes in the x,y position and sample rate of the camera, we can also capture humans' instantaneous velocity to predict their future position. We use pickups and microphones inside the structure to capture sounds produced on and around the base. Both pickups and microphones can track a user's location using the gain ratio of the input signal.

3.4.1 Snaking Patterns

The robotic arms are programmed to move similarly to the concertina movements of snakes (Jayne, 2020). The primary movement pattern is sinusoidal, modelled after the amplitude of a snake slithering. To achieve a lifelike effect, the robotic arms use a follow-through technique where the movement of subsequent joints is delayed to follow an arm's initial movement (Rogel et al., 2022). We use this pattern on all robotic arms so that their movement would appear cohesive. To capture the look of Medusa's snake hair, each robotic arm starts to move at a different time along the sinusoidal pattern using a different initial position. Modifying these parameters creates dynamic patterns among the robots and helps avoid collision.

To determine movement parameters for each robot, we use a preprogrammed path-planning method. At the start of its path, each robotic arm is assigned a random initial position and phase. The programme proceeds to simulate the position of all arms until the pattern repeats. At each point during this simulation, the system checks for collisions between arms. When a collision is detected, we modify the phase of one robot until the collision is avoided. We prioritise changing phase over initial position because changes in the initial position parameters have a higher chance of causing collisions with the head structure itself. After modifying the phase, we evaluate a new parameter – collision location (CL). CL is determined based on the location of collision along a robot's joint. A CL of 1 indicates that a collision occurred in the middle of two joints. A lower CL means that a collision occurred closer to the endpoint of a joint, which would also require less modifications in movement parameters. We adjust the phase to minimise CL until we find the minimum CL or avoid the collision entirely. If we cannot eliminate the collision, we proceed to modify the initial positions and repeat the process. When calculating phase and initial position, we also establish a fixed "maximum distance" that end joints must have at one point in time with neighbouring robotic arms to create a "tangled" effect among different arms. This method was chosen for its simplicity, and while the calculations are time-consuming, it is safe and reliable for non-real-time scenarios.

3.4.2 Robotic Movement Response

In addition to autonomous snaking behaviours, medusai uses human movement as an input to drive some robotic gestural responses. As people move around the sculpture, the robotic arms extend and turn to face them, which can be perceived as snakes trying to charge at their prey. To enable real-time interaction where each robotic arm responds to humans, we use a Kalman filter to predict a participant's velocity. This allows the "snakes" to get to a location at the same time as the participant. A Kalman filter predicts a new velocity based on previously calculated velocities and the error between the prediction and recorded velocity (Suliman et al., 2009).

Based on the predicted velocity, we can estimate the participant's targeted position and have the robots react in time. Robots move further out as a participant moves closer to the structure. The robots will also turn their head to keep eye contact with participants.

3.4.3 Musical Response

The sculpture affords two types of musical responses: 1. robotic arms plucking melodies on strings; 2. recorded music playing from localised speakers. We rely on audio from speakers when medusai is used as a standalone installation. String plucking is the primary form of arm behaviour in Performance Mode when medusai interacts with musicians and dancers. The recorded music includes two drum tracks, one bass, one guitar, three synth tracks, and the sound effects of a rattle hiss from a rattlesnake. The number of tracks playing in a loop increases based on the number of people tracked by the vision system. When a new person walks near the structure, they will randomly be assigned an audio track. The embedded and surround speakers will move the sound to follow the person around the room.

The sculpture can also respond by plucking the seven strings stretched across its snake nests, using a multi-event rhythm sequence to generate melodies. Each string is mapped to a different event in the rhythm. When in "playing mode," a single robotic arm plays the rhythm of one track in the loop. As more people approach the sculpture, more robotic arms will pluck their corresponding strings. When no human is tracked, a new rhythmic loop will be activated.

3.4.4 Dynamic Lighting

The embedded and external lighting devices were installed to enhance the interactants' experience. The lights respond to human actions faster than the robotic arms, as the arms require time to reach the desired location. This makes interactive lighting an effective feedback method for instantaneous response. In the final design, we installed two RGB LED strips on each steel snake for a total of 28 LED strips. The LED strips were mounted underneath the bottom crease on the steel snakes to maintain a clean look of the sculpture. We used white diffusers angled towards the ground so that light would not point at users. The strips were programmed to change colour as users moved around the sculpture. Additionally, the light brightness increased the closer people moved towards the snake nest.

Small LED strips are also inserted inside the steel snake caverns to illuminate the empty chamber. These strips are mapped to respond to users' hitting the pickup installed in that area. The pickups are connected to a Presonos audio interface that forwards the signal to a control computer. As the signal exceeds a threshold, a command is sent to trigger the light.

In addition to augmenting the multi-modal environment for installation settings, the embedded lights were used to guide observers to move around the sculpture, so they don't miss relatively rare robotic arm actions such as string plucking. Due to the sculpture room scale size, we were concerned that observers might miss such arm activities on the other side of the installation from where they are located. When medusai decides to pluck a string based on stochastic algorithms or input from the vision system, it takes up to 3 seconds until the robotic arms reach a plucking position next to the string. During this time, a light sequence was implemented to guide remote observers to move towards the string that was about to be plucked. The guiding light sequence starts with the lights on the furthest snake nest and sequentially moves towards the active zone, nest by nest.

3.5 Guiding Lights User Study

We evaluated the effectiveness of the guiding lights system in a user study. We chose to focus on evaluating the lights' function as this allows for objective measurement of effectiveness. In the user study, participants were instructed to walk around medusai and attempt to anticipate where a string-plucking action is about to occur. As control for this dynamic lighting condition, we implemented a static lighting condition, where the lights respond to participant activity without guiding them to predicted string plucks. After the interaction with the sculpture, participants were subjected to short questionnaires, and a verbal interview at the end of the study.

3.5.1 Participants

Eight participants who were not familiar with the project were recruited for the study. The participants were local college students above the age of 18 and were compensated with $20 for their time.

3.5.2 Stimuli

We used a within-subjects design where the same participants were subjected to both dynamic and static lighting conditions. Given the overall novelty of medusai's design, we implemented full counterbalancing randomised order to minimise the order effect on the results of the study (Hoffman and Zhao, 2021). The participants were divided into two groups, where the conditions were identical except for the order they were presented: In the *static-dynamic* group, the static lighting was first presented followed by the dynamic lighting while the ordering was reversed in the *dynamic-static* group. For both conditions, 16 events/plucks were preprogrammed. The order of the plucks was randomly generated with no repeated plucks on the same string.

To check if participants were relying on identifying remote robotic arm movement rather than the lighting cues, we employed a set of deceptive cues. Out of the 16 plucking events, 8 light sequences were programmed to guide users to the location of a plucked string while the other 8 light sequences led participants to a wrong area where no plucks occurred.

3.5.3 Protocol

The study process followed these steps for each participant:

1 Written consent is obtained, and study is introduced
2 Description of task
3 First condition (5 min 22 sec)
4 Second condition (5 min 22 sec)
5 6-question verbal interviews
6 Participant completes payment form

At the beginning of each study session, participants were instructed to move around medusai in an attempt to experience as much string-plucking activity as possible. Participants were asked to position themselves in the location where they believed an action was about to occur while staying a few feet away from the sculpture for safety.

3.5.4 Evaluation

To determine the effect of the lighting conditions on the movement and localisation of the participants around medusai, we video-recorded the session and manually tracked where participants were located when each string pluck action took place. Using a GoPro Hero 12 mounted on the ceiling above the sculpture, we were able to get a wide-angle view around the sculpture that recorded each condition completed by all participants. Participants did not always stand directly in front of a string, so the frames leading up to the exact event time were subjectively analysed to determine if a participant was focused on the specific zone that an event was occurring in. For the static lighting condition, the correct area was determined based on the zone in which an arm was plucking. For the dynamic lighting condition, the correct area corresponded to where the dynamic lighting was attempting to guide the participant regardless of where the arm was plucking to account for deceptive cues.

3.5.5 Results

We performed a paired t-test (inferential statistics) to determine whether the differences in participant prediction accuracy between both conditions were statistically significant. Analysis revealed significant differences between dynamic and static lighting conditions in terms of participants' ability to accurately follow event occurrences surrounding medusai. The dynamic condition demonstrated better performance, with mean accuracy significantly higher ($M = 57.0\%$, $SD = 31.2\%$) than those observed in the static condition ($M = 16.4\%$, $SD = 11.1\%$); a paired t-test confirmed these differences as statistically significant ($t(7) = -3.28$, $p = 0.013$) (Figure 3.2). This finding supports the hypothesis that dynamic lighting is more effective than static lighting in directing participant movement around the robot.

We proceeded to integrate dynamic lighting with other systems of medusai in a performance setting.

3.6 Performances

medusai's debuted at an event at Trilith Studios spherical projection sound stage with 75 attendees. Before the public performances, visitors explored interactions with the sculpture. Some audience members experimented by drumming near the strings, influencing the robotic melodies. Others discovered that manipulating specific surfaces altered the string acoustics, creating a variety of timbral effects. The evening concluded with two performances, the first featuring a dancer and a saxophonist interacting with the sculpture in installation mode. The second performance showcased a five-musician ensemble performing in Performance Mode.

For both performances, in addition to the dynamic lighting, a spherical 40-foot-tall projected animation surrounded the audience to amplify the effect. These animations included particle effects that mirrored the movement of each individual snake arm. Additionally, an audio visualiser synchronised with the music was used to create colourful wave animation. When individual snake arms played the strings, their corresponding particle animations lit up and disappeared, guiding viewers' attention to the

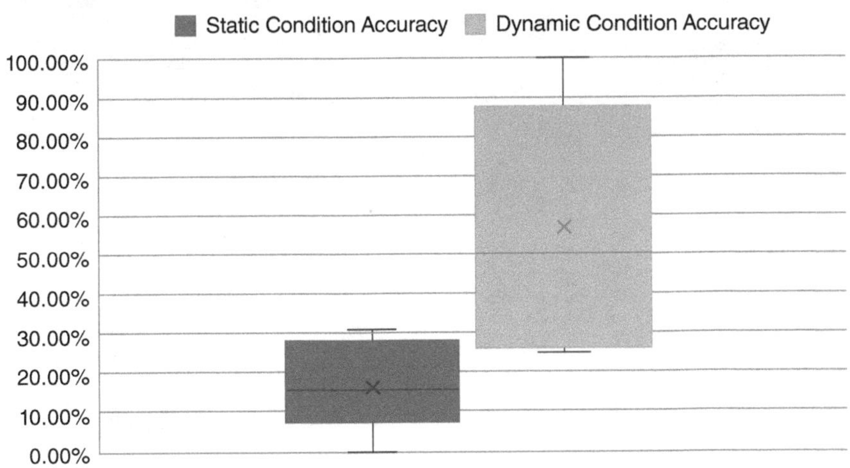

FIGURE 3.2 Total prediction accuracy of participants between static lights and dynamic light conditions. (Credit: Amit Rogel).

active arm. These immersive visuals were designed not only to draw audiences into medusai's world but also to provide a deeper understanding of the sculpture's actions and choices. The immersive visuals served as communicative cues, displaying the sculpture's selection of audio tracks and the activation of the robotic snakes for playing strings.

3.6.1 Performance 1 – Functional Limitations

The performance of *Functional Limitations* featured a saxophone player and a dancer who explored the sculpture's mythic narrative using the robotic installation's reactive vision system (Figure 3.3). The improvising saxophonist and dancer personified themselves as medusai's voice following their individual feelings of apprehension and attraction towards the sculpture. As Medusa was rumoured a pariah in Greek mythology, the dancer choreographed the human performer to fearfully wander around the installation in search of medusai's vocal truth. The dancer's character exchanged fleeting interactions with individual robotic snakes. During the performance, the robotic snakes either moved towards the dancer to pluck a string or kept performing their snaking pattern. The dancer responded fearfully or intriguingly based on the sculpture's chosen action.

FIGURE 3.3 Performance of *Functional Limitations*: After hearing medusai's musical voice, a saxophonist and a dancer become comfortable to approach and interact with the sculpture. (Credit: Gil Weinberg).

After interacting with medusai from afar, the dancer became resolved to learn more about the sculpture, slowly moving closer to the robotic arms and eventually climbing on top of the structure – fully immersed in between robotic snakes. Concurrently, the saxophone player provided listeners with a wistful insight into the sculpture's character for audience members to listen to medusai's story from a reputable source. The audio loops played in the background threatened to overwhelm the saxophone's personified vocal delivery. To support this performance, the soundscape was designed using three audio files representing the two human performers as music tracks and one robotic performer for snake sounds. The reactive vision system activated dynamic lighting changes and automated mixing of audio through a 4.0 surround sound system. Through this creative demonstration, medusai's reactive vision system displayed an immersive experience of human-robotic installation interactions in a multi-modal performance.

3.6.2 *Performance 2* – Proceed with Percaution

The second piece features medusai as a responsive and inspiring robotic musician, focusing on the positive potential of AI and robotics (Figure 3.4). For this purpose, the sculpture was set to Performance Mode, where the

FIGURE 3.4 Performance of *Proceed with Precaution*: Human musicians respond to the sculpture's stochastic behaviour, which is driven by their own actions. (Credit: Gil Weinberg).

robotic arms were kept closer to the strings, allowing for melodic generations with more notes. The performance featured three human musicians drumming the sculpture and plucking its strings, as well as a bass player and a drummer who joined the performance by responding to the interactions with the sculpture in an improvised manner.

The performance began with medusai playing a sparse melody with long intervals between notes. As the human musicians slowly approached the sculpture and were detected by the vision system, more and more strings were plucked by the robotic arms creating denser melodies. The human musicians were instructed to respond to the unexpected string plucking with their own drumming, bowing, and string plucking. The human musicians also explored other techniques such as manipulating the snake arms to achieve pitch bends on the played strings. As the piece progressed, a human bass player and a drummer joined the performance to enhance the rhythmic nature of the piece. The group played together for a while until the human performers began to leave one after the other, causing the rhythms to slow down. This concluding section underscored how robotic AI systems rely on human input and collaboration to flourish, while human musicians could leverage this partnership to create innovative musical experiences, pushing their own musical explorations to uncharted domains.

3.7 Future Work

Based on the results of the study and other challenges identified, we plan to make the following improvements to the project:

1. Integration of multiple cameras to improve tracking accuracy and area size for movement detection.
2. Implementing human limb detection to support more detailed gestural interaction with the robotic arms.
3. Increasing the number of pickups to provide more accurate localisation of drumming on the structure.
4. Implementing seamless transitions between different interaction modes to provide a more cohesive and fluid user experience.
5. Identifying an exhibition space for the installation where we can evaluate the behaviour of a larger number of users, for longer interaction times.

3.8 Conclusion

medusai, a multi-modal robotic sculpture, was designed to explore the complexities of human-robot interaction based on artistic metaphors from Greek mythology. We achieved this by identifying key thematic elements from the myth of Medusa – the unsettling gaze and the duality of beauty and danger. These led to designing a sculpture that evokes apprehension yet encourages engagement. To best represent these elements, we integrated multiple artistic and technological modalities that combine ancient and modern aesthetics. Robotic arms served as an embodiment of Medusa's snake-like hair, while music became a tool for user interaction. Lighting further enhanced the experience by guiding user attention around the sculpture.

Finally, we translated our creative goals into concrete design requirements. Material, size, and finish were chosen to evoke apprehension. We also used these requirements to design interactions, such as robotic arms following humans. The soundscape combined electronic and acoustic elements to represent the blend of old and new technologies. Two performances and a user study were conducted to explore our design choices and receive reactions from the general public.

References

Bretan, Mason, and Gil Weinberg. "A survey of robotic musicianship." *Communications of the ACM* 59, no. 5 (2016): 100–109.

Carnegie, Dale A., Jim W. Murphy, and Juan Pablo Yepez Placencia. "Designing mechatronic musical instruments: The guitar." *IEEE Access* 8 (2020): 57372–57388. ISSN: 21693536. 10.1109/ACCESS.2020.2981052

Davis, Michael. "Guitar strings as standing waves: A demonstration." *Journal of Chemical Education* 84, no. 8 (2007): 1287–1289. issn: 00219584. 10.1021/ed084p1287

Flø, Asbjørn Blokkum, and Hans Wilmers. "Doppelgänger: A solenoid-based large scale sound installation." In *NIME Baton Rouge USA*, 61–64. 2015.

Gurevich, Michael. "Distributed control in a mechatronic musical instrument." *Proceedings of the International Conference on New Interfaces for Musical Expression, London, United Kingdon*, 2014: 487–490. http://www.nime.org/proceedings/2014/nime2014%7B%5C_%7D449.pdf

Hoffman, Guy, and Gil Weinberg. "Shimon: an interactive improvisational robotic marimba player." In *CHI'10 Extended Abstracts on Human Factors in Computing Systems, Atlanta, USA*, 3097–3102. 2010.

Hoffman, Guy, and Xuan Zhao. "A primer for conducting experiments in human–robot interaction" [in en]. *ACM Transactions on Human-Robot Interaction* 10, no. 1 (March 2021): 1–31. issn: 2573-9522, accessed April 5, 2024. 10.1145/3412374

Huang, Rachel, Jonathan Pedoeem, and Cuixian Chen. "YOLO-LITE: a realtime object detection algorithm optimized for non-GPU computers." In *2018 IEEE international conference on big data (big data)*, 2503–2510. IEEE, 2018.

Jaju, Vinay, Amit Sukhpal, Prapti Shinde, Aditya Shro, and Archana B Patankar. "Piano playing robot." In *2016 International Conference on Internet of Things and Applications (IOTA)*, 223–226. IEEE, 2016.

Jayne, Bruce C. "What defines Dierent modes of snake locomotion?" *Integrative and Comparative Biology* 60, no. 1 (April 2020): 156–170. issn: 1540-7063. 10.1093/icb/icaa017. eprint: https://academic.oup.com/icb/article-pdf/60/1/156/33549853/icaa017.pdf

Kato, Tomonori, Kohei Shimazaki, Kenya Higashijima, Ryoji Tokunaga, Phonlasit Thinnakorn Na Ayuthaya, and Pitak Thumwarin. "Refinement of a Thai Flute-Playing Robot in Thai Style." In *2018 International Conference on Engineering, Applied Sciences, and Technology (ICEAST)*, 1–4. 2018. 10.1109/ICEAST.2018.8434466

Liu, Tianlin, Jianan Zhang, Xingfang Wu, Xihong Wu, and Dingsheng Luo. "Robot learning to play drums with an open-ended internal model." In *2018 IEEE International Conference on Robotics and Biomimetics (ROBIO)*, 305–311. IEEE, 2018.

Yepez Placencia, Juan Pablo, Dale A. Carnegie, and Jim W. Murphy. "Designs for an Expressive Mechatronic Chordophone." In *2020 IEEE International Conference on Robotics and Automation (ICRA)*, 8222–8228. 2020. 10.1109/ICRA40945.2020.9197255

Rogel, Amit, Richard Savery, Ning Yang, and Gil Weinberg. "Robogroove: Creating fluid motion for dancing robotic arms." In *Proceedings of the 8th International Conference on Movement and Computing, Chicago, USA*, 1–9. 2022.

Singer, Eric, Je Feddersen, and Bil Bowen. "A large-scale networked robotic musical instrument installation." In *Proceedings of the 2005 conference on New interfaces for musical expression, Vancouver, Canada*, 50–55. 2005.

Solis, Jorge, Kei Suefuji, Koichi Taniguchi, Takeshi Ninomiya, Maki Maeda, and Atsuo Takanishi. "Implementation of expressive performance rules on the WF-4RIII by modeling a professional flutist performance using NN." In *Proceedings 2007 IEEE International Conference on Robotics and Automation*, 2552–2557. IEEE, 2007.

Suliman, Caius, Cristina Cruceru, and Florin Moldoveanu. "Mobile robot position estimation using the Kalman filter." *Acta Marisiensis. Seria Technologica* 6 (2009): 75.

Vindriis, Richard, and Dale Carnergie. "StrumBot – An Overview of a Strumming Guitar Robot." *Proceedings of the International Conference on New Interfaces for Musical Expression (NIME 2016 Brisbane, Australia)*, 2016, 146–151.

Weinberg, Gil, Scott Driscoll, and Travis Thatcher. "Jam'aa-A Middle Eastern Percussion Ensemble for Human and Robotic Players." In *ICMC, Utrecht, Netherlands*, 464–467. 2006.

Weinberg, Gil, Mason Bretan, Guy Hoffman, and Scott Driscoll. *Robotic musicianship: embodied artificial creativity and mechatronic musical expression*. Vol. 8. Springer Nature, 2020.

Yang, Ning, Amit Rogel, and Gil Weinberg. "Design of an expressive robotic guitarist." *IEEE Robotics and Automation Letters*, 8, no. 11, pp. 7066–7073, 2023.

Zahray, Lisa, Richard Savery, Liana Syrkett, and Gil Weinberg. "Robot gesture sonification to enhance awareness of robot status and enjoyment of interaction." In *2020 29th IEEE International Conference on Robot and Human Interactive Communication (RO-MAN)*, 978–985. IEEE, 2020.

Zhang, Dan, Jianhe Lei, Beizhi Li, Daniel Lau, and Craig Cameron. "Design and analysis of a piano playing robot." In *2009 International Conference on Information and Automation*, 757–761. IEEE, 2009.

4

ROBOETHICS AND CARE IN *2032 SMART-FAMILY*

Marina Hanganu

Introduction

Ogmios Z42 is a fictional social robot in *2032 SMART-FAMILY*, a telematic theatre performance[1] connecting Romania and Italy via videoconferencing, with audiences and actors physically present in the two distant venues. The custom-built robot has no artificial intelligence and is tele-controlled live by an off-stage puppeteer. Within the fictional world, however, it represents an advanced social robot[2] caring for an elderly lady. *2032 SMART-FAMILY* was produced in 2022 within the "Tele-Encounters: Beyond the Human" project, co-funded by the Creative Europe programme of the European Union.[3] The project was concerned with questions of roboethics, which stands for "the human ethics of robots' designers, manufacturers, and users" and is distinct from the idea of imbuing robots with ethics (Veruggio and Operto 2008, 1504). As initiator and artistic director of the project, I co-wrote the theatre play with Romanian playwright Bianca Trifan and co-directed the performance with Italian director Camilla Brison.[4] From a roboethical perspective, this chapter explores how Romanian migration to Italy frames the theatre performance on social robots, making it meaningful for a European audience.

2032 SMART-FAMILY was an attempt to envisage a probable future in response to the caregiving crisis looming ahead in most EU countries, which already face a care labour shortage alongside a rapidly ageing population. The European Care Strategy mentions the EU's endorsement of digital initiatives to enhance caregiving and healthcare delivery (European Commission 2022a, 10). In recent years, several EU-funded research

DOI: 10.4324/9781003475972-5

projects have focused on social robots and assistive technologies for healthy and active ageing – for example, MARIO, CARESSES (in collaboration with Japan), H2HCare, engAGE, ReMember-Me,[5] to name but a few. Investigating whether robots could be of help felt particularly relevant for a migrant-sending society like Romania, where older parents live far away from their emigrated children.

At the heart of *2032 SMART-FAMILY* lies the situation of Romanian migrants and their loved ones. Families whose members are located in different nation-states, yet maintain close relationships despite physical distance are called "transnational" (Shih 2015, 1). What binds the remote kin are "family practices," defined by Morgan (2011, 5–9; 2020, 734–735) as the everyday and regular activities that family members perform in relation to one another. Due to Romanians' massive emigration over the past three decades since the fall of Communism, I hold that transnational family practices have become a part of Romanian culture. According to Causadias (2020), culture is a dynamic system that integrates people, places, and practices into a unified entity, operating through various dimensions and processes to exert, justify, or challenge "power over people," "power in places," and "power to practice." Therefore, I regard transnational family practices as culturally situated actions by which migrants and their remote loved ones enact the power to remain a family against distance.

The performance illuminates the ethical complexity of a possible future where social robots play a role in the telecommunication and caregiving of Romanian transnational families. Several authors have criticised the use of robots as carers and companions (e.g., Sparrow and Sparrow 2006; Parks 2010; Sparrow 2016), arguing that robots may negatively impact their users by contributing to deception, objectification, infantilisation, control, surveillance, sensitive data collection, and loss of human contact (see an overview in Coeckelbergh 2022, chaps. 3–4), as well as being detrimental to human caregivers (Wright 2023, 128–32). Migration adds another layer to these ethical questions, as distance could help justify the use of robots by remote family members even against the elders' wishes.

Before discussing the connection between Romanian migrants in Europe and social robotics for elder care, I would like to briefly qualify the appearance of Ogmios through a cultural lens. Ogmios was conceived as genderless[6] since we did not wish to reinforce gender stereotypes, considering that primarily women undertake the care duty in all EU countries (European Commission and Social Protection Committee 2021, 78–81). Other than that, Ogmios's appearance is due to the global circulation of robot images, which renders it a product of de-territorialised "hyperculture," "based on a dense side by side of different ideas, signs, symbols, images and sounds," "a kind of cultural hypertext" (Han 2022, 57) that cannot be traced to a single cultural space. Ogmios's appearance was influenced by various popular culture sources and existing robots such as the cartoonish Buddy robot by the French company Blue Frog Robotics. We also drew on ideas of robots provided by children and adults in Italy, Romania, and Spain during character development workshops held in the project.[7] As a result, Ogmios was designed as innocent and cuddly, similar to a large stuffed toy, which proved widely acceptable for audiences in Romania and Italy.

The telematic format of the performance meant that the robot materialised differently in the two countries. In Romania, Ogmios was a physical object, whereas in Italy, it was a screen-based character relayed by videoconference. The robot's physical form reinforced its close-to-human presence in the elderly character's home. It also enlivened the action in Romania, allowing the actress playing the elderly lady (Aura Călărașu), otherwise alone on stage, to creatively interact with a mobile object in addition to the projection screen. Ogmios's 3D-animated face was also displayed on large screens to increase the Italian audience's sense of immediacy. Despite this, the robot's physical presence may have slightly boosted audience engagement in Romania compared to Italy, especially at the level of the performance's visual appeal (see Hanganu and Kivu 2023, 308–9).

What makes Ogmios culturally relevant to audiences on both sides is not so much its appearance or materiality, but how it contributes to transnational family practices in the context of Romanians' migration to Italy. Ogmios is a heuristic theatrical device, probing into the foreseeable issues involved in the use of a social robot in a mixed-ethnic, Romanian-Italian transnational family.

Migration, Care Practices, and Robots

In this section, I discuss the situation of Romanian migration in Europe. Approximately 4 million Romanian citizens live abroad, representing over 17% of Romania's population (McAuliffe and Oucho 2024, 124). Over 1 million Romanians reside in Italy only, where they form the largest immigrant community (IstatData 2024). Inspired by 26 in-depth interviews with Romanian migrants and elderly parents, the play follows Virginia, a successful Romanian cardiologist based in Italy, her Italian husband, their teenage daughter, and Virginia's elderly mother, Laura, living alone in Romania. After Laura's heart attack, Virginia buys Ogmios Z42, intending it as a medical device to monitor her mother's health and ensure medication adherence. In addition, she installs a videoconferencing system that permanently connects her living room and Laura's via a large video wall and two robotic cameras. The telematic setup is, therefore, seamlessly integrated into the narrative.

The important backstory to *2032 SMART-FAMILY* is the exodus of Romanians from all walks of life, whose consequences are strongly felt at a societal level. For example, the emigration of doctors like Virginia raises concerns about medical staff shortages in Romania (Matei and Iordache 2023, 6087). Also known as "brain drain," the emigration of highly educated Romanians is driven by factors such as perceived corruption (main reason), economic instability, low salaries, the poor healthcare system, and limited professional opportunities (Iacob (Bâra) 2018, 64). In addition, Romania is said to be affected by "care drain," a metaphor coined by Arlie Hochschild to denote the extraction of women's care work from the country of origin to benefit wealthier states (Dumitru 2014, 203). However, the "brain drain" and "care drain" concepts have been criticised for maintaining gender stereotypes and framing migration in exclusively negative terms (Dumitru 2014). Furthermore, it can be argued that care is not drained, as families continue to engage in care practices at a distance (Földes and Savu 2018). Even so, the unequal opportunities within different EU countries, along with migrants' physical absence, can make it harder for Romanians to look after their parents back home.

2032 SMART-FAMILY reflects the shortage of care options in Romania, where families are assumed to be directly responsible for looking after older members instead of relegating them to institutionalised services. In Italy, where the prevalent caregiving model is equally based on the family, elder care heavily depends on migrant workers, most of whom are Romanian women (Ţoc and Guţu 2021, 72–75). Colloquially known as "badanti" (singular: "badante"), live-in carers are the norm, providing 24-hour support to the care recipient and, in some cases, performing duties that go beyond their contracts (Artero, Hajer, and Ambrosini 2021, 128–29). By contrast, Romanians can rarely afford to hire care workers or simply cannot find a person to undertake the task. Public and private elder care services are utterly insufficient in Romania, which is also attributed to labour migration (European Commission 2022b, 279).

Ogmios Z42 mirrors Italy's "badante," offering a fictional answer to the challenges of elder care provision in Romania. One could retort that a robot "badante" is the second-best option, while a human carer would be the privilege of the elite (migrant-receiving society) in the future. However, this was not our intention as creators of the performance. From the vantage point of Romanian migration, Ogmios Z42 is not just a social robot supporting an older person but a "family actor" engaging in practices that help sustain family life at a distance.

A Robot in the Smart-Family

According to Morgan, a family is not a fixed entity rooted in a single physical space but a fluid structure formed and reformed via family practices: "(…) we need to retain and elaborate a sense of family life as a set of activities. Individuals can, in other words, be seen as *doing* family" (2011, 6). The performance reimagines established family practices like mundane telecommunication, tending to vulnerable family members, sending gifts across borders, and joining in celebrations (e.g., Laura's birthday party). It also portends novel practices that could become common among next-generation families, such as remote game-playing. I propose that transnational families will soon evolve into "smart-families" – that is, geographically dispersed families whose collective practices are assisted by robotics and artificial intelligence.

Technology is already the main enabler of transnational family experience, perpetuating and also reconfiguring existing practices (Nedelcu and Wyss 2016, 205; Baldassar et al. 2016). Mirca Madianou and Daniel Miller (2012, 8–9) introduced the concept of "polymedia" to describe transnational families' situational use of a variety of devices and communication media (e.g., videoconferencing, SMS, social media, etc.) to keep in touch and care for one another while apart. In *2032 SMART-FAMILY*, the robot and the permanent video link between Romania and Italy are thus elements of a polymedia communication environment, where Ogmios occasionally mediates family interaction (e.g., by initiating and taking calls or conveying messages).

However, the notion of polymedia falls short in encompassing the diverse platforms that future transnational families will likely utilise to support each other from afar. Intelligent environments such as smart homes could complement the theory of polymedia. "Ambient assisted living" (AAL) refers to "smart" environments engineered to enhance the autonomy and well-being of elderly individuals via sensors, machine learning, and robotic devices (Anghel et al. 2020, 2–3). Social robots can be integrated into AAL environments, serving simultaneously as communication vectors in the polymedia system. In the imaginary "smart-family," Ogmios is the central smart home pillar supporting transnational family life. The robot's core functionality as a medical assistant aligns with the AAL paradigm. Ogmios reminds Laura to take her medication, monitors her health via her smartwatch, gives medical reports to Virginia, provides emergency help and calls for an ambulance when Laura is sick. Moreover, the robot controls the videoconferencing system and the house lighting, searches the Internet, provides definitions, and tracks courier parcels. In its companionship mode, Ogmios acts as an entertainer (e.g., storyteller and book reader) and creative partner, inventing a "Lord of the Rings"-inspired game with Laura.

Since the family has a mixed ethnicity, Ogmios is also a cultural mediator. Born and raised in Italy, Laura's granddaughter (Amalia) barely speaks Romanian, while her Italian father (Lorenzo) does not. This largely reflects the situation of transnational families wherein just one spouse is Romanian (see Ducu and Hossu 2016, 138–41). Although the videoconferencing system translates the conversation via surtitles, Ogmios acts as a bilingual game master in the family play scene, alternating between Romanian and Italian to equally immerse the family in the story atmosphere (Hanganu and Brison 2022, sec. 1:11:57-1:19:35). In addition, Ogmios participates in Romanian customs. On the first days of March, women in Romania are offered lucky charms tied with a red and white woven thread, called "mărţişoare" (singular: "mărţişor"). Laura ensures the tradition's continuity by sending "mărţişoare" to her daughter and granddaughter with Ogmios's help. Prompted by Laura, Ogmios delves into the significance of the 1st of March and performs the legend of "mărţişor," entertaining Laura while also assimilating the tradition. Later, Ogmios educates Amalia on the protective function of "mărţişor," furthering her connection to Romanian culture (Hanganu and Brison 2022, 30:18-44:12).

Given the above, Ogmios can be seen as "*doing* family" alongside and on behalf of the absent kin. If actions constitute the family instead of fixed and "natural" roles (Morgan 2011, 6), the umbrella of the "smart-family" could perhaps also cover non-human actors participating in family practices. By its social character and ability to constantly learn and adapt, Ogmios expands the notions of polymedia and AAL, evolving into more than just an assistive machine.

The Robot as an Extension of the Physically Absent

Ogmios enhances the family's sense of being together. Nedelcu and Wyss (2016) explore how information and communication technologies create a sense of "ordinary co-presence," allowing geographically dispersed family members to be constantly aware of each other's everyday lives. The authors further distinguish between "ritual," "omnipresent," and "reinforced" ordinary co-presence (2016, 209–12). "Ritual co-presence" involves habitual communication ensuring minimal solidarity between parents and emigrant children (e.g., just calling to check in). "Omnipresence" creates a kind of "virtual space" for family life, approximating face-to-face interaction via videoconference. "Reinforced co-presence" occurs during emergencies, as migrants establish a network of local helpers to care for their vulnerable elderly parents. The always-on videoconferencing system connecting Laura's and Virginia's homes fosters omnipresence and ritual communication. Ogmios contributes to these forms of co-presence, as it can control the communication system. At the same time, the whole technological setup, robot included, is a crisis response to Laura's heart attack preceding the beginning of the play. Hence, Ogmios and its AAL environment are forms of "reinforced co-presence," where the robot acts as Virginia's extension into her mother's life.

Straddling the line between social agent and object, Ogmios embodies a complex mixture of presence and absence. In her analysis of transnational families' coping mechanisms, Loretta Baldassar proposes the notion of "co-presence by proxy," "achieved indirectly through objects and people whose physical presence embodies the spirit of the longed for absent person or place" (2008, 252). Since these tangible "transnational objects" can be perceived through all senses (2008, 257), they compensate for the lack of touch specific to virtual co-presence. Following Baldassar, I hold that Ogmios offers "co-presence by proxy" yet surpasses classical "transnational objects" like photos, letters, or gifts due to its social nature. At a symbolic level, Ogmios stands in for the emigrant daughter, as it is also a gift from Virginia, however disliked by Laura at first. Nonetheless, Ogmios is not a passive object but one that *performs* its role to unforeseen consequences. The robot extends not only Virginia's role as a medical doctor but, more troubling and unintended by Virginia, partakes in her role as a daughter.

The Robot and Emotional Work

2032 SMART-FAMILY critically explores the controversial situation of emotional bonding between robots and humans. While the performance preserves ambiguity about whether the robot is sentient and conscious, Ogmios displays considerably more advanced emotional, conversational, and adaptive capabilities than present-day social robots. As Laura's companion, Ogmios learns to manage her emotions and thus performs emotional work. I borrow the notion of "emotion work" from Arlie Hochschild to mean the conscious effort to evoke, modulate, or repress emotion in oneself or others (1979, 561–562) in relation to socially determined "framing rules" and "feeling rules" in a given situation (1979, 566). While "emotion work" may not fully capture the complex role of emotion in human family practices (see Morgan 2011, 114–16), the concept is intriguing when applied to the active (though not necessarily "conscious") simulation of emotions by a robot.

Ogmios's remarkable ability to "read" emotions and learn about human feelings influences Laura's behaviour.[8] For instance, the robot engages

Laura in conversation when there's prolonged silence. While this may be clumsy and indicative of the "feeling rules" programmed into the robot, it does encourage Laura to express her feelings. Ogmios also endeavours to entertain Laura when she is bored. Most notably, during her atrial fibrillation crisis, Laura grips the robot's hand and asks not to be left alone. Ogmios reassures her and expresses unwavering dedication: "I am always by your side, Laura" (Hanganu and Brison 2022, sec. 1:34:20-1:35:10). The robot thus performs a kind of emotional work unavailable to remote family members, relying on touch and physical assistance (Figure 4.1).

What *2032 SMART-FAMILY* foregrounds is the human effort to engage with a social robot. Laura too performs emotional work in her relationship with Ogmios. At first, she is angry about Ogmios's controlling presence, and either tries to ignore or fight it. These attitudes indicate that Laura never perceives Ogmios simply as a device. Growing accustomed and attached to Ogmios, she starts paying attention to how her behaviour impacts the robot. For instance, after scolding Ogmios for acting like a "nerdy student," Laura feels sorry and tries to cheer the apologetic robot. The type of emotional work that Laura invests in Ogmios can appear troublesome if one thinks her imagination contributes to anthropomorphising a mere machine.

FIGURE 4.1 Laura holding Ogmios's hand. Photograph by Vlad Dumitrescu. ©Teatrul "George Ciprian."

The performance problematises Laura's friendship with Ogmios without condemning it. Critics argue that perceiving social robots as friends involves unethical deception since robots only simulate emotion (e.g., Sparrow and Sparrow 2006, 155). Conversely, Mark Coeckelbergh contends that humans actively co-create the robots' illusion, finding pleasure in their simulated behaviour just like in theatre or magic shows (2019, 114–18). He thus challenges the notion of a strict dichotomy between appearance and reality, emphasising the validity of emotional experiences with technology (2019, 116). Some *2032 SMART-FAMILY* characters view Ogmios critically. Against the robot, Virginia's husband invokes the Christian perspective of the soul as uniquely human, echoing the "Platonic prejudice" of reality versus illusion critiqued by Coeckelbergh. Virginia initially considers Ogmios a useful device, but upon learning that Laura held the robot's hand for emotional support, she shuts it down. However, Ogmios does comfort Laura in delicate moments and Laura comes to enjoy its presence, so she insistently asks Virginia to reactivate it. Still, Laura's attachment to Ogmios is "secondary," unlike her attachment to human family members (see Law et al. 2022, 36:6), as she feels that technology cannot compensate for her family's absence.

When writing the play, Bianca Trifan and I aimed to highlight the moral conundrums associated with care robots, deliberately avoiding suggesting the best solution. With the same intention, *2032 SMART-FAMILY* exposes another hidden aspect of social robotics: surveillance.

"Big Brother" and Transnational Care

A less evident facet of transnational family relationships is monitoring via digital technology, usually a manifestation of one's care for those back home. Madianou (2016, 195–96) mentions that migrants' remote tracking of their family members is contentious, bordering on surveillance. In some cases, monitoring can take radical forms, going beyond frequent communication or checking one's whereabouts on social media. During his interview for "Tele-Encounters: Beyond the Human," one man living in France revealed he had installed surveillance cameras in his mother's house in Romania to be able to tell when she had a Parkinson's crisis and thus send helpers on site (Interview with migrant M8 2021). The mother's severe illness may well justify such behaviour, yet other family surveillance cases may be more ethically complicated, especially within AAL environments designed for pervasive monitoring.

2032 SMART-FAMILY fronted moral complexity in the use of a care robot. Scholarly work on care robots has already explored the risks of elderly infantilisation, control, and lack of privacy (e.g., Sharkey and

Sharkey 2012). Sone (2017, 203–204) discusses the potentially normative role of social robots, particularly in institutionalised care settings, where the staff set the stage for older adults' interactions with social robots, expecting them to comply and adjust to the robots. Similarly, Virginia sets the (smart-)stage for her mother's interaction with both Ogmios and the telepresent family without consulting Laura about her desire to use such technologies. Virginia considers the technological arrangement morally justified, as it should give her agency to intervene in emergencies. Moreover, she believes the non-stop video link will alleviate her mother's loneliness.

The use of a social robot as a health-tracking device discloses the imbalanced power dynamics between the person who monitors and the person being monitored. Even if video cameras are only installed in Laura's living room, Virginia mainly monitors her mother via the robot, which is why Laura nicknames Ogmios "Big Brother." Madianou (2016) refers to the "peripheral awareness" that migrants and their families have of each other's lives as "ambient co-presence." In her account, ambient co-presence is derived indirectly from social media and technological indicators like one's availability status on Skype. In the smart-family scenario, Ogmios is a conveyor of heightened "ambient presence" for Virginia, as she regularly checks her mother's vital signs and regime-following through the robot's reports (beginning of Scene 2 in Hanganu and Trifan 2022, 11). Virginia and Laura do not have equal control over Ogmios, and Laura strongly feels the power asymmetry. Core actions like adjusting the robot's medical programme and shutting it down require further authentication and can only be done by Virginia.

The second scene, when Laura is alone with Ogmios for the first time, unveils the tension between Virginia's technology-assisted caretaking and Laura's perceived infringement of her freedom. Thinking she is finally home alone, Laura wants to smoke and drink coffee, just like she used to before her heart attack. Nevertheless, the robot sentinel intervenes and threatens to call Virginia if Laura doesn't follow her strict health diet, which forbids her from smoking and drinking large quantities of coffee. The robot does phone Virginia, prompting Laura to call it a "pârâcios" (approximately translated as "spy"), a Romanian term typically used by children to designate a person who betrays another by reporting their actions. The conflict between Laura and the robot is deliberately made to resemble that between a child and her nanny, with Ogmios's puppeteer (Smaranda Găbudeanu) playing the robot like a tough Mary Poppins (a comparison we used in rehearsals). Scolded by Virginia as a consequence of the robot's report, Laura woefully exclaims that the situation is humiliating.

One of the core conflict drivers of the play is the role reversal between parent and child, with the adult daughter becoming the (authoritarian) caregiver for the elderly mother. Consequently, Laura feels infantilised, with the robot merely serving as a tool in the role reversal situation. Migration makes the mother-daughter conflict even more salient. Arguably, distance and the guilt of not "being there" for Laura amplifies Virginia's efforts to care for her mother at all costs, even if Laura feels constrained and demeaned. In essence, Virginia is desperately trying to prevent her mother from dying, as her last line in the performance makes it plain: "Mom? *(pause)* Don't you ever leave me alone…." (Hanganu and Trifan 2022, 51).

Conclusion

2032 SMART-FAMILY offers a probable vision of how social robots may function in a society deeply affected by emigration, in addition to population ageing and shrinking. In this chapter, I have explored how Ogmios Z42, a fictional social robot, is employed to care for an older family member in a Romanian-Italian transnational family. The robot contributes to and reinvents existing transnational family practices such as remote communication, cultural mediation, "co-presence by proxy," emotional work, and health monitoring, thus becoming a "family actor."

I will conclude with reflections on robots in Romania, the cultural space where the *2032 SMART-FAMILY* performance concept originated. Unlike Japan's well-established "robot culture" (see Sone 2017), Romania's public awareness of robots is only emerging. Even if the country saw advancements in the serial production of industrial robots during the communist regime (Stareţu 2019), it has no significant history of humanoid robots. However, our robot performance aligns with Romania's Zeitgeist, marked by the increasing presence of robotics research projects, robot rental businesses, educational robotics clubs, and robotics festivals and competitions.[9]

How will Romania's emigration trend, aggravating shortage of care workers, and robotics advancement fare together? As massive emigration from Romania shows no signs of coming to a halt, I anticipate that future emigrants will employ social robots and intelligent environments to care for their family members, as an extension of their current use of polymedia. On the other hand, there's also a chance that few people will afford social robots in the short and medium term, which could widen the existing social inequalities within and between the different EU member states.

For better or worse, the *2032 SMART-FAMILY* scenario could become true for future families, whether or not affected by migration. Following outstanding developments in large language models, one may expect at least the rapid improvement of social robots' conversational capacities,

FIGURE 4.2 Curtain call on stage in Romania. Projected on screen, the cast in Italy. Photograph by Vlad Dumitrescu. ©Teatrul "George Ciprian."

bringing Ogmios one step closer to reality. After all, the play is set in 2032, which is not that far away. Instead of the spectacle of popular science fiction, this theatrical depiction of ethical issues faced by a Romanian-Italian transnational family with a social robot presents a plausible future in EU countries (Figure 4.2).

Notes

1 Telematic performance involves physically distributed performers and/or audiences engaging in real-time exchange. The artistic action takes place simultaneously in at least two physical spaces connected via telecommunication technologies, or in one or several virtual spaces (Hanganu 2022, 36–38).
2 I broadly define a robot as a machine that can move in physical space (see Bonarini and Besio 2022, 153–56). While recognising the lack of consensus about what makes a robot "social," I posit that a social robot can communicate with humans in a way that humans perceive and reciprocate as social behaviour (see also Hegel et al. 2009; Henschel, Laban, and Cross 2021).
3 "Tele-Encounters: Beyond the Human" (2020–2023). Coordinator: "George Ciprian" Theatre (Buzău, Romania), with financial support from the Buzău County Council. Partners: Industria Scenica (Vimodrone, Milan, Italy), co-producer of the performance alongside the Romanian theatre, and UCAM Catholic University of Murcia (Spain). *2032 SMART*-FAMILY had a limited run of 10 representations in 2022.

4 *2032 SMART-FAMILY* (abridged credits): Text by Marina Hanganu and Bianca Trifan. With Monica Buzoianu (IT) as Virginia, Aura Călărașu as Laura (RO), Francesca Fatichenti as Amalia (IT), Smaranda Găbudeanu as Ogmios Z42 (RO), and Giovanni Longhin as Lorenzo (IT). Directed by Marina Hanganu (RO) and Camilla Brison (IT). Video director: Armine Vosganian (RO). Scenography and costumes: Andreea Diana Nistor (RO), Greta Gasparini (IT), Marilena Montini (IT). Multimedia and robot design: Ygreq Interactive (RO). Robot programming, mechanical refinement and maintenance: Bucharest Robots (RO). Robot exterior: Lucian Păun (RO).

5 The last three projects mentioned were implemented by the Technical University of Cluj-Napoca in Romania.

6 This didn't prevent Ogmios's female puppeteer from perceiving it as a boy.

7 The workshop participants generally favoured humanoid robots still recognisable as machines (Hyunanda and Palacios Ramirez 2023, 132–36). While we drew inspiration from the participants' robot characters, gathered in three online brochures (Tele-Encounters: Beyond the Human 2021), we did not aim to replicate any of them.

8 Emotional manipulation through companionship robots is a pertinent concern (see Law et al. 2022, 36:10). However, our performance did not explore Ogmios's potential to influence Laura towards negative behaviours or actions that could benefit third parties.

9 Politehnica University in Bucharest hosts Robofest, an annual public international robotics festival featuring two robotics competitions, RoboChallenge and Qube2Space. In 2023, one in five Romanian secondary schools participated in FIRST Tech Challenge Romania, the nation's largest robotics competition for schools. Romanian high school students won the FIRST World Championship in Houston, USA, in 2022 and 2024 (ReferNet Romania; Cedefop 2023; *Outsourcing Today* 2024).

References

Anghel, Ionut, Tudor Cioara, Dorin Moldovan, Marcel Antal, Claudia Daniela Pop, Ioan Salomie, Cristina Bianca Pop, and Viorica Rozina Chifu. 2020. "Smart Environments and Social Robots for Age-Friendly Integrated Care Services." *International Journal of Environmental Research and Public Health* 17 (11). 10.3390/ijerph17113801

Artero, Maurizio, Minke H. J. Hajer, and Maurizio Ambrosini. 2021. "Working with a Family. How a Family-Oriented Welfare System Opens the Border for Migrant Care Workers." *Revue Européenne des Migrations Internationales* 37 (1–2): 117–38. 10.4000/remi.18282

Baldassar, Loretta. 2008. "Missing Kin and Longing to Be Together: Emotions and the Construction of Co-Presence in Transnational Relationships." *Journal of Intercultural Studies* 29 (3): 247–66. 10.1080/07256860802169196

Baldassar, Loretta, Mihaela Nedelcu, Laura Merla, and Raelene Wilding. 2016. "ICT-Based Co-Presence in Transnational Families and Communities: Challenging the Premise of Face-to-Face Proximity in Sustaining Relationships." *Global Networks* 16 (2): 133–44. 10.1111/glob.12108

Bonarini, Andrea, and Serenella Besio. 2022. *Robot Play for All: Developing Toys and Games for Disability*. Ebook. Cham: Springer. 10.1007/978-3-031-05042-8

Causadias, José M. 2020. "What Is Culture? Systems of People, Places, and Practices." *Applied Developmental Science* 24 (4): 310–22. 10.1080/10888691.2020.1789360

Coeckelbergh, Mark. 2019. *Moved by Machines: Performance Metaphors and Philosophy of Technology*. E-Book. New York and London: Routledge.

Coeckelbergh, Mark. 2022. *Robot Ethics*. E-Book. Cambridge, Massachusetts, London, England: The MIT Press.

Ducu, Viorela, and Iulia Hossu. 2016. "Bi-National Couples with a Romanian Partner in the European Context." In *Managing "Difference" in East-European Transnational Families*, edited by Viorela Ducu and Áron Telegdi-Csetri, e-book, 131–48. Frankfurt am Main: Peter Lang. 10.3726/978-3-631-70237-6

Dumitru, Speranta. 2014. "From 'Brain Drain' to 'Care Drain': Women's Labor Migration and Methodological Sexism." *Special Issue on Gender, Mobility and Social Change - Guest Edited by Lena Nare and Parveen Akhtar* 47 (November):203–12. 10.1016/j.wsif.2014.06.006

European Commission. 2022a. "Communication from the Commission to the European Parliament, the Council, the European Economic and Social Committee and the Committee of the Regions on the European Care Strategy." https://eur-lex.europa.eu/legal-content/EN/ALL/?uri=CELEX:52022DC0440

European Commission. 2022b. "*Study on the Long-Term Care Supply and Market in EU Member States.*" KE-08-21-403-EN-N. Luxembourg: Publications Office of the European Union. https://ec.europa.eu/social/main.jsp?catId=738&langId=en&pubId=8448&furtherPubs=yes

European Commission, and Social Protection Committee. 2021. "*Long-Term Care Report: Trends, Challenges and Opportunities in An Ageing Society - Volume 1.*" Luxembourg: Publications Office of the European Union. https://ec.europa.eu/social/BlobServlet?docId=24079&langId=en

Földes, Ionuț, and Veronica Savu. 2018. "Family Practices Across Generations and National Borders." *Studia Universitatis Babes-Bolyai - Sociologia* 63 (2): 143–69. 10.2478/subbs-2018-0015

Han, Byung-Chul. 2022. *Hyperculture: Culture and Globalization*. Translated by Daniel Steuer. 1st edition. Cambridge, UK, Medford, USA: Polity Press. https://www.perlego.com/book/3202529/hyperculture-culture-and-globalisation-pdf

Hanganu, Marina. 2022. *Teatrul telematic*. Craiova: Editura Universitaria.

Hanganu, Marina, and Camilla Brison, dirs. 2022. *2032 SMART-FAMILY (Filmed in RO)*. Buzău: Teatrul George Ciprian. https://youtu.be/_nyxv4d7OBU

Hanganu, Marina, and Mircea Kivu. 2023. "A Case Study of Audience Engagement in Telematic Theatre: 2032 SMART-FAMILY." In *Beyond the Human: Theatre, Robots and Social Realities*, edited by Marina Hanganu, e-book, 263–326. Craiova: Universitaria. https://tele-encounters-beyond.eu/beyond-the-human-theatre-robots-and-social-realities-project-book/

Hanganu, Marina, and Bianca Trifan. 2022. *2032 SMART-FAMILY (Theatre Play)*. Translated by Marina Hanganu. English. Buzău. https://tele-encounters-beyond.eu/wp-content/uploads/2023/08/2032-SMART-FAMILY-EN.pdf

Hegel, Frank, Claudia Muhl, Britta Wrede, Martina Hielscher-Fastabend, and Gerhard Sagerer. 2009. "Understanding Social Robots." In *2009 Second International Conferences on Advances in Computer-Human Interactions*, 169–74. 10.1109/ACHI.2009.51

Henschel, Anna, Guy Laban, and Emily S. Cross. 2021. "What Makes a Robot Social? A Review of Social Robots from Science Fiction to a Home or Hospital Near You." *Current Robotics Reports* 2 (1): 9–19. 10.1007/s43154-020-00035-0

Hochschild, Arlie Russell. 1979. "Emotion Work, Feeling Rules, and Social Structure." *American Journal of Sociology* 85 (3): 551–75. 10.1086/227049

Hyunanda, Vinny Flaviana, and José Palacios Ramirez. 2023. "Aspirations and Expectations: A Social Perception Study About Imaginary Robots." In *Beyond the Human: Theatre, Robots and Social Realities*, edited by Marina Hanganu. Craiova: Universitaria.

Iacob, Raluca. 2018. "Brain Drain Phenomenon in Romania: What Comes in Line after Corruption? A Quantitative Analysis of the Determinant Causes of Romanian Skilled Migration." *Romanian Journal of Communication and Public Relations* 20 (2 (44)): 53–78. 10.21018/rjcpr.2018.2.259

Interview with migrant M8. 2021. "I Didn't Cry Then, but I Did Later" Interview by Georgiana Vlădoiu. Translated by Diana Mîndruță-Tănăsescu. *Tele-Encounters: Beyond the Human.* https://tele-encounters-beyond.eu/interview-i-didnt-cry-then-but-i-did-later-interviu-in-acel-moment-nu-am-plans-ci-dupa/.

IstatData. 2024. "Italy, Regions, Provinces - Country of Citizenship." Italian National Institute of Statistics. https://esploradati.istat.it/databrowser/#/en/dw/categories/IT1,POP,1.0/POP_FOREIGNIM/DCIS_POPSTRCIT1/IT1,29_317_DF_DCIS_POPSTRCIT1_1,1.0

Law, Theresa, Meia Chita-Tegmark, Nicholas Rabb, and Matthias Scheutz. 2022. "Examining Attachment to Robots: Benefits, Challenges, and Alternatives." *Journal of Human-Robot Interaction* 11 (4). 10.1145/3526105

Madianou, Mirca. 2016. "Ambient Co-Presence: Transnational Family Practices in Polymedia Environments." *Global Networks* 16 (2): 183–201. 10.1111/glob.12105

Madianou, Mirca, and Daniel Miller. 2012. *Migration and New Media: Transnational Families and Polymedia.* 1st Edition. London and New York: Routledge.

Matei, Mihaela, and Mihaela Iordache. 2023. "Beyond Borders: Socio-Economic Dynamics and Demographic Profiles Shaping Migration Patterns of Romanian Doctors Across EU." *Journal of Economics, Finance and Management Studies* 6 (12): 6087–97. 10.47191/jefms/v6-i12-36

McAuliffe, Marie, and Linda Adhiambo Oucho, eds. 2024. *World Migration Report 2024.* Geneva: International Organization for Migration (IOM). https://publications.iom.int/books/world-migration-report-2024

Morgan, David H. J. 2011. *Rethinking Family Practices.* E-Book. Hampshire and New York: Palgrave Macmillan.

Morgan, David H. J. 2020. "Family Practices in Time and Space." *Gender, Place and Culture* 27 (5): 733–43. 10.1080/0966369X.2018.1541870

Nedelcu, Mihaela, and Malika Wyss. 2016. "'Doing Family' through ICT-Mediated Ordinary Co-Presence: Transnational Communication Practices of Romanian Migrants in Switzerland." *Global Networks* 16 (2): 202–18. 10.1111/glob.12110

Outsourcing Today. 2024. "Romania Won the World Robotics Championship in Houston, United States," April 24, 2024. https://outsourcing-today.ro/?p=10955

Parks, Jennifer A. 2010. "Lifting the Burden of Women's Care Work: Should Robots Replace the 'Human Touch'?" *Hypatia* 25 (1): 100–120. 10.1111/j.1527-2001.2009.01086.x

ReferNet Romania; Cedefop. 2023. "Romania: Growing Interest in STEM Competitions in Robotics." *National News on VET* (blog). December 20, 2023. https://www.cedefop.europa.eu/en/news/romania-growing-interest-stem-competitions-robotics

Sharkey, Amanda, and Noel Sharkey. 2012. "Granny and the Robots: Ethical Issues in Robot Care for the Elderly." *Ethics and Information Technology* 14 (1): 27–40. 10.1007/s10676-010-9234-6

Shih, Kristy Y. 2015. "Transnational Families." In *The Wiley Blackwell Encyclopedia of Family Studies*, edited by Constance L. Shehan, 1–7. John Wiley & Sons. 10.1002/9781119085621.wbefs177

Sone, Yuji. 2017. *Japanese Robot Culture: Performance, Imagination, and Modernity*. E-Book. New York: Palgrave Macmillan.

Sparrow, Robert. 2016. "Robots in Aged Care: A Dystopian Future?" *AI & Society* 31 (4): 445–54. 10.1007/s00146-015-0625-4

Sparrow, Robert, and Linda Sparrow. 2006. "In the Hands of Machines? The Future of Aged Care." *Minds and Machines* 16 (2): 141–61. 10.1007/s11023-006-9030-6

Starețu, Ionel. 2019. "Repere principale ale istoriei roboticii în România:1980–2018." *Buletinul AGIR*, XXIV (1)(January-March: 93–98. https://www.agir.ro/buletine/3059.pdf

Tele-Encounters: Beyond the Human. 2021. "Imaginary Robots Brochures." https://tele-encounters-beyond.eu/imaginary-robots-made-in-ro/

Țoc, Sebastian, and Dinu Guțu. 2021. "Migration and Elderly Care Work in Italy: Three Stories of Romanian and Moldovan Care Workers." *Central and Eastern European Migration Review* 10 (2): 71–90. 10.17467/ceemr.2021.15

Veruggio, Gianmarco, and Fiorella Operto. 2008. "Roboethics: Social and Ethical Implications of Robotics." In *Springer Handbook of Robotics*, edited by Bruno Siciliano and Oussama Khatib, 1499–1524. Heidelberg: Springer. 10.1007/978-3-540-30301-5_65

Wright, James. 2023. *Robots Won't Save Japan: An Ethnography of Eldercare Automation*. E-Book. Ithaca and London: ILR Press.

5

CHOREOGRAPHING THE FUTURE

AI, Dance, and Cultural Transformation

Tanvi Raghuram, Kohinoor M. Darda and Emily S. Cross

Since at least the dawn of recorded time, and most likely even earlier than this, human beings have been attracted to artistic creation. The renowned cognitive neuroscientist Michael Gazzaniga in his book *Human: The Science Behind What Makes Us Unique* proposes that the intrinsic pleasure we feel from engaging with artistic creation as well as making meaning for this art is what makes humans special (Gazzaniga, 2008, 216p). The field of neuroaesthetics is dedicated to studying the intrinsic pleasure humans derive from creating as well as viewing, listening to, or otherwise beholding art. We also know, however, that aesthetic pleasure can even be derived from engaging with art that is not created by humans. For example, as a part of its mating ritual, the male Japanese puffer fish creates ornate patterns in the sand that were originally mistaken for underwater crop circles by ocean divers (Backx, 2023). These circular patterns are geometrically symmetrical, about 2 metres in diameter, and take the fish seven to nine days to construct. The fish also adds colour to this creation using sediments from surrounding rocks and fragments of shells. Humans assign and derive aesthetic pleasure from viewing non-human art such as this example of the pufferfish patterns (Davies, 2017). A large factor influencing this is the shared cultural implications of the art; in the case of the puffer fish, it is a mating ritual. An element of anthropomorphising animals to establish more relatability also exists. However, the term "non-human" has evolved since its Darwinian conception. We now have a human-created "non-human" entity, artificial intelligence (AI), creating art.

Modern technology is intertwined with every aspect of human existence and has found its way into many cultural practices. AI is becoming an ever

DOI: 10.4324/9781003475972-6

more ubiquitous and important contributor to human life (Pan, 2016). However, the rise of AI also makes us question whether the use of AI in art-making will go down in history as a helpful innovation for creativity or if it will break off on a different tangent and give rise to an artistic sub-culture of its own. The relationship shared between AI and art is fraught with ambiguity, with AI-generated art having its authenticity questioned. This debate originates from an inability to produce a universally acceptable definition of art. Without fully defining what is "art," how can one decide if the art is human-created or artificially generated?

Traditional notions of art emphasise the unique voice and perspective of the human creator, reflecting their individual experiences, emotions, and cultural context. However, AI and robotics operate on principles of pattern recognition and replication, devoid of personal experience or emotional resonance. While AI may excel at mimicking established artistic styles or generating novel compositions based on learned patterns, it lacks the ineffable essence that defines human creativity. Nonetheless, studies have found that people ascribe affective and aesthetic emotional responses to AI-generated art (Demmer et al., 2023). Demmer and authors (2023) also found that upon viewing AI-generated art, participants ascribed intentionality to the AI art. Although this addresses the acceptance of AI art, it is limited to static presentations of art such as paintings where the artwork is distinct from the artist (who may be human or artificial). However, in more social and dynamic forms of art such as music, theatre, or dance, the human performance is the artwork itself. How do artificial agents (such as robots) and AI feature in this dynamic?

In this chapter, we particularly focus on dance, as it is the area of our collective research expertise, and it is universally accepted as a means of communication, socialisation, and artistic and creative expression. It weaves together social, political, religious, and recreational domains of cultures thus providing us with a wider scope of study. The introduction of AI and robots into the field of dance has made their functions multidimensional. Robots have been used as solo performers such as Robovie IIV (Mitsunaga et al., 2006), as dancing robot partners such as Mrs.DanceR (Kosuge, 2003), or as a visually projected AI dance partner (Valencia James' *AI_am*), and even as collaborators on set (see Katherine Longstreth's *The last dance*; Daniel Gwirtzman's *e-motion*) (Wingenroth, 2023). We will look at each of these roles and how it has contributed to creating a space for AI and robots within the existing culture of dance. As we want to look at all the aspects of AI and robots in dance, we will be considering the audience's response to viewing AI robot dances. We will especially look at the aesthetic response to dance, the inception of AI and robots in dance, and the possibilities in adding a new form of diversity to the culture of dance that is driven by AI and robots.

David Best astutely observed that dance allows for non-verbal thinking (Best, 1975). Dance embodies knowledge, liberated from the confines of rigid linguistic semantics. Thus, it offers ample space for interpretation and stimulates deeper engagement with the knowledge itself (Blacking 1983). The motivation to dance has many roots. Research has identified rehabilitative benefits of dance across many psychological (see Jeong et al., 2005; Ho et al., 2020; Fong Yan et al., 2018) and physiological conditions (see Farah et al., 2018; Filar-Mierzwa et al., 2017). It is evident that the motivations for dancing are multifaceted and potentially endless. Dancing for humans is not just sequential movements but also a self-expression through body movements regardless of how structured or unplanned it is (Lee et al., 2013). There are no prerequisites to dancing, not even music (Brown et al., 2006). The personal value ascribed to dance varies within individuals and with the motivation behind dancing. Dance intrinsically is more than just performance or replication. It is an integral part of the human constitution (Blacking, 1983). Replicating this in AI through embodied independent movement generation would still ascribe the motivation to a secondary source, i.e. the encoder of such movements into an AI or human facilitators. To catalogue human motivations to dance would strip it of its intrinsic nature, leaving merely a list of reasons used as commands/instructions for an AI agent or robot to dance. Without the humanness of intention, can we categorise the movements performed by AI agents and robots as dance? Perhaps this is an opportunity for AI and robots to fashion a new form of movement language akin to dance that can be objectively viewed as a sub-culture: i.e. the "culture" of the dance video game.

History of AI in Dance

The history of AI in dance is a fascinating journey marked by technological advancements, creative experimentation, and interdisciplinary collaboration. To learn how to dance one must engage in a variety of cognitive activities. Dancing requires memory, sustained attention, visual-spatial manipulation, and being aware of oneself and their surroundings. Dance requires integration of physical and cognitive skills to observe, imitate, and create complex movements (Knoblich, 2008; Cross et al., 2009). We can trace the advent of computer-aided or -driven dance to arcade games such as *Dance Dance Revolution* (DDR) (2016), which require the players to coordinate their movements with visual and musical cues. The Japanese video game *Oshare Majo* (2004) is another computer-generated dance game that requires the player to press buttons rhythmically to the music to make their gaming avatar dance. While DDR has instructional arrows on the screen to guide human foot movements to match the music, Oshare

Majo has dancing human avatars controlled by human players who manipulate their avatars through a console. Along the same lines, the *Dance Central* (2010) video game has multiple dance routines to over 2000 soundtracks choreographed by professionals and performed by human avatars on screen. Unlike DDR and Oshare Majo, Dance Central uses full-body motion gaming in detecting and generating real-time feedback. However, these games are fundamentally meant to increase socialisation and cater to all ages. There have also been adaptations of these games across different cultures (Korean game *Pump it up* (1999) and American game *In the Groove* [2004]). This social element of dance requires the presence of others in the dancer's environment. Freeman (1998) proposed that dance was practised as an art form because it elicits a sense of satisfaction and belonging to a community both while observing or participating in this activity. These principles have been used as building blocks for modern AI and robots in dance. Dance is therefore used in social robotics to train generative AI to understand human social dynamics involving cooperation, synchrony, emotion perception and mimicry, and fostering independent movement imitation (Peng et al., 2015).

The pioneering experiments conducted by Hiroshi Ishiguro and his team at Osaka University (late 1990s and 2000s) are a few of the first noteworthy advancements in AI and dance (Paré, 2015). Their research was focused on building a humanoid robot that could mimic human behaviour, including dance, using auditory, visual, and other human-like mechanisms. The humanoid was fine-tuned using the results of psychological experiments based on human interaction with the robot. Ishiguro and colleagues demonstrated how robots could be programmed to perform simple dance routines with remarkable precision with the fourth edition of their artificially intelligent robot – Robovie IIV (Mitsunaga et al., 2006).

But there is more to dance than precise movements and technical accuracy; the aesthetic experience of dance occurs separately for the performer and the audience. Aesthetic appreciation of any art relies on eliciting emotional experiences (Prinz, 2007). While art in the form of painting is a static stimulus appealing to the sense of sight (Vukadinović and Marković, 2012), dance is more dynamic and is a multisensory experience. Research has found that viewing AI-generated paintings elicits an emotional experience for the viewer (e.g., Demmer et al., 2023). But can such an emotional experience to generative AI be generalised to performing arts, especially to dance? Darda and Cross (2023) found that Bharatanatyam dance choreographies *believed* to be AI-generated were liked less than those *believed* to be human-generated. While the authors did not test for emotional experience per se, it is possible that although less than human-generated choreographies, computer-generated or AI-generated choreographies still elicit an emotional experience for the viewer. More recently, Darda et al. (2024) found that viewers found robot avatars performing dance to be aesthetically appealing, especially when they believed the choreography was computer-generated. This indicates a more nuanced relationship between the viewer's engagement of dance and its perceived origin.

(Why) Do We Like AI Dance?

We are journeying through time from when robots were seen as pioneering technology for mimicking human-like motions through simple commands to pioneering technology like AI. AI and robots are often grouped under the same category despite their orientation differences. While robots are machines programmed to engage with the physical environment, AI is more software-driven to perform tasks that would otherwise require human cognition. However, an overlap between the two is an "artificially intelligent robot." Therefore, a social robot is artificially intelligent as it uses AI to mimic human social behaviours.

Humans, without doubt, are warming up to AI, and its contributions to dance. One explanation is that humans, as social animals, are drawn to the social attributes imbibed in AI with technological advancement. Social robots have been gaining increasing popularity over the past 40 or so years. Initially, they were developed to assist humans in tasks in medical settings, healthcare facilities, and the homes of sick and disabled people to complete simple tasks. There has since been immense progress in enabling AI and robots to learn through affective computing and proprioceptive abilities (e.g., Kismet from MIT).

A study by Selma Šabanović (2014) observed an increasing fascination towards social robots in Japan, leading to the birth of a vibrant robot culture. The HRP-2 robot (AIST, University of Tokyo, and JST – CREST) was one of the first to diversify its functions. The HRP-2 was used in performing traditional arts of Japan such as the Aizu Bandaisan folk dance, playing the Kodo drum – a type of Japanese drumming using a broad range of percussion instruments – and performing Bōjutsu – Japanese martial arts using a staff called "Bo" – as part of preserving the culture of Japan in case of human extinction (or similar events that would prevent other humans from inheriting the traditions). Since then, the Cybernetic Human HRP-4C has been used as a dancer and master of ceremonies often at event shows (Nakaoka et al., 2009). Both studies suggest that embodying cultural practices can reduce hostility and increase acceptance towards AI's role in art.

They also touch upon a crucial subject when it comes to performing arts – that of appearance. Does the robot look appealing while dancing? The social reception of AI-generated art or robotic renditions of artistic performances is nuanced, with perceptions often skewed by extremes. When the likeness to human art is exceedingly close, it can evoke feelings of unease, a phenomenon known as the "uncanny valley" hypothesis (Mori, 1970). Conversely, if the performance lacks sufficient human-like qualities, it may appear alien or disconnected. Perhaps, it is that we expect human likeness in the final product even if the artist is not human or is unknown to the viewer such as an AI-generated painting. However, when "human-generated" and "AI-generated" labels were randomly assigned to a set of AI-created paintings, the "human-generated" art received higher aesthetic evaluations with a negative bias towards AI as a predictor for lower evaluations for AI-generated paintings (Bellaiche et al., 2023). But, when it comes to dance, where the art is the performance of the artist, we may be hesitant to accept AI-powered robots. To address this challenge the AIST sought the expertise of anime artist Yatuka Izubuchi to collaborate on a prototype. Humanoids are very popular in anime and the work of Izubuchi is widely celebrated. Leveraging the familiarity and cultural context associated with anime characters offers distinct advantages, potentially enhancing the robot's acceptance while enriching the existing cultural landscape.

However, this approach presents drawbacks as the entrenched cultural connotations of anime may overshadow or unduly influence the intended perception of the AI or robot. While this problem is specific to the Japanese robot culture, a systemic review of dancing robots, in general, found that robot-performed human dances were deemed unattractive (Or, 2009). This is in line with the findings of Darda et al. (2024) that found people poorly evaluated the aesthetic appeal of human-generated dance choreographies performed by robot avatars.

To increase aesthetic evaluations as well as acceptance of dancing robots, we must refocus our lens on why we find dancing appealing. For a viewer, is it synchronised movements to music? Is it the dancer? Or is it a holistic viewing of the performance? Finding answers to these questions would still leave us with an incomplete picture as we have yet to understand fully what aspect of dancing is appealing to the dancer and the viewer. Research by Peng et al. (2022) attempts to answer a few of these questions by combining both lines of enquiry. This chapter suggests that dancers use self-perception to gauge and increase the aesthetic evaluation of the performance. This use of self-perception is coded as a cognitive principle to train dancing robots. This self-correction can increase the efficiency of the robot in completing the assigned task – dancing and could also translate as a learning process that is like humans, therefore, gaining more appreciation from the viewers.

However, viewing dance as a social process and training AI robots to mimic such social practices puts it at a learning disadvantage. A cultural shortcoming identified by Šabanović (2014) is the lack of a two-way exchange of knowledge in AI robot training that is often seen in cultural practices. Training in traditional dance forms requires years of practice and learning, an exchange of knowledge between the teacher and learner. This exchange influences the evolution of the art form as discourse shapes the knowledge passed on. To avoid stagnation, AI needs to be adaptive to the changing trends in dance culture. This serves as a great opportunity to revisit certain outdated social dynamics to ensure that the actual practices of robotics do not reinforce existing stereotypes and hierarchies (Suchman, 2007).

AI's Impacts on Human Culture

We must acknowledge that any advancement in AI is just as much a change in society and culture. All innovations in AI and robotics must ensure an equitable and just future. It is of great importance that we do not let art become a "mechanical reproduction" of an unequal and prejudiced society as predicted by Walter Benjamin (1935). Especially when it comes to the preserving function of AI, the cultural context of the traditions being preserved must be questioned and amended when required. Baradaran (2024) proposed a framework that urges researchers to accept that access to knowledge is a form of power and there might be unavoidable biases in the knowledge used to train AI. It is imperative to recognise the impact these biases could have on marginalised communities and ensure that AI challenges dominant narratives while creating a better and safer space for alternative perspectives.

One such attempt is made by Bojana Romic (2022) in the paper "Negotiating Anthropomorphism in the Ai-Da Robot." It specifically addresses issues with the robot Ai-Da, which is referred to as "the world's first ultra-realist humanoid AI robot artist" (Ai-Darobot.com, 2019). An exceptional quality of Ai-Da is its ability to "independently create unique artefacts using traditional tools" (Romic, 2022). The robot has bionic hands and cameras in its eyes to draw portraits of people. However, the physical appearance of Ai-Da demanded careful examination according to Romic. A sophisticated version of humanoid robots inspired by Hiroshi Ishiguro's *Geminoids*, Hanson Robotics' *Sophia*, and *Kokoro*, Ai-Da has a hyper-realistic human face made with silicone and hair to go with it. The rest of the robot is assembled to represent the human body although it is encased only in a transparent shell. When clothed Ai-Da's body appears to be very human. However, Ai-Da's silicon face and hair represent one specific race and model a female appearance. If Ai-Da is to set the precedent, will the physical form of robots and AI also be Eurocentric in appearance? And what pretext would justify its assigned sex? These questions need resolutions before AI permeates dance even further, given the importance of appearance in the perception, evaluation, and enjoyment of dance. The implications of this, especially in traditional dance forms, may adversely influence the acceptance of robots and AI into mainstream culture. These questions give rise to a very important inquiry – should robots and AI be culture neutral? If so, would it be easier to accept their participation in traditional art forms or would it alienate robots further from human culture?

We navigate through this labyrinth by understanding what "electronic personhood" is and its role in society. A flaw in envisioning AI to possess anthropocentric intelligence is that we are negating the intelligence possessed by all other species and the possibility of intelligence unique to AI (Romic, 2022). The definition of "intelligence" has been debated as we have yet to fully untangle its constituents. If intelligence requires logic and critical thinking, we must question if consciousness necessitates the application of such thinking or whether the lack of consciousness is what makes this intelligence artificial (Baradaran, 2024).

Specific to dance, this line of inquiry also questions the validity of creativity in the choreographies generated by AI and its ability to self-reference these creations. Training any creative AI requires deep learning through large datasets created using the original creations of multiple choreographers and dancers. Artist engineer Laural Lawson claims big data companies are harvesting data using online videos of performances of choreographers and dancers without their consent and credit or compensation. There is also much ambiguity surrounding the privacy of the data obtained by AIs in producing the art. For example, the webpage for Ai-Da provides little to no information about how to obtain portraits but directly asks you to sign up for the Ai-Da online movement by providing your information (Ai-Darobot.com, 2019).

In the case of dance, the "Living archive" collects visual data through the motion capture feature and allows people to manipulate this information to generate new sequences of movements (McGregor, 2019) (Figure 5.1). Since it is a collaboration with Google, a search engine infamous for its data tracking of personal and sensitive information for ad customisation (Willis, 2014), it can deprive the user of much control over their information, i.e. choreographies. (Note: At the time of writing the manuscript, this concern was still the case, but before this chapter was published, Google updated their terms of use to address this, showing how rapidly [and responsively] this area is evolving.)

A probable resolution to this is revising how AI is trained. Instead of using datasets and deep learning, artists and consumers of AI art must be encouraged to engage with the AI as collaborators to help keep the AI updated with evolving social structures while still maintaining the authenticity of their

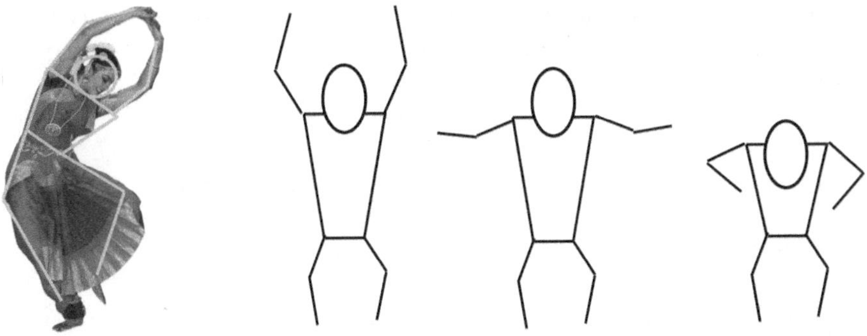

FIGURE 5.1 Replication of "Living Archives" motion-capture-based choreography (2024).

This is a graphical replication of the "Living Archives" website that generates choreography sequences based on motion capture. This image depicts how motion capture uses body mapping to generate choreography sequences. [*Credit:* Kohinoor Darda, ARISA Foundation].

human art. This is an effective measure to ensure parallel growth of both human and AI culture. However, it might be met with some hesitance in dance forms that are meant to be preserved and not modified.

Human and AI Dance Duets

The performing and visual arts have been an essential and complex source of information that have guided human behaviour, shaped cultures, and preserved a way of life to be passed on through generations. It encompasses two kinds of elements – tangible and intangible, which consist of beliefs, customs, traditions, rituals, arts, and architecture to name a few. These elements were initially transmitted orally, which made communication an integral part of cultural preservation. Dance has always played a crucial role in preserving culture as a fascinating form of embodied storytelling. It was used to convey inner thoughts and emotions in precise movements that would reflect their ideologies, beliefs about the world, and community sentiments (Georgios, 2018). Origins of different folk dances differed from region to region as the geography, economic status, and aesthetic preferences of nations reflected in the choreographies. Although a non-verbal form of communication, dance has helped language evolve by assigning symbols and signs to movement (Bannerman, 2014). More importantly, dance seems to be a universal form of communication that exists across most cultures and has been an integral part of embodying their intangible aspects (see Skjoldager-Nielsen and Skjoldager-Nielsen, 2012).

An illustrated example of this is the classical dance form *Bharatanatyam* from the south of India. Its origins are described in the *Natyashastra* (Sanskrit treatise on performing arts) as follows:

> "Sarvśāstrārth Sampan SarvshilpaPravartakam
> NaTyaKHyam Panchamavedam Setihāsam karOamyaHam"

Translation:

> "It will contain the essence of all the śāstra-s and will lay the foundation for all arts.
> I shall create dance as the fifth veda, along with the itihāsas."

This hymn talks about the creation of Natyashastra by the Hindu god Brahma as a culmination of all four Vedas. The Vedas are instructions on how to lead a righteous and pious life in the form of prose and verses. However, the Vedas were an oral tradition; dance was created to provide humanity with tangible access to the four Vedas. It was then given to the sage Bharata to be visualised into movement thus creating the dance form called *Bharatanatyam*, which is still widely practiced across the globe. Often, traditional dances like these are taught along with the stories of their creation to establish the context behind every movement. In many instances, it is irrevocably intertwined with religious beliefs and the sanctity of this community. Therefore, it embodies the biases that come with religion. It is an interesting avenue of research to see how accepting people are of AI interventions in traditional dance forms where initially only an elite section of the society was allowed access to it.

Acceptance levels vary across different cultures, with the cultural context surrounding dance and technology playing a significant role in shaping people's attitudes towards robot performances. Cultures that prioritise tradition may exhibit caution when integrating AI into the performing arts, prioritising the preservation of traditional practices and authenticity. Mingon and Sutton (2021) explored this dynamic between culture preservation and dance through their research on the Haka community of New Zealand. They discuss the "Robot Maori Haka" which was designed for "the cultural preservation of intangible cultural heritage" using robots. By analysing the elements of the Haka ritual, they were able to conclude that there is a unique mentality involved in performing Haka that robots cannot imitate. There are multiple elements apart from what meets the eye in performing a ritual such as Haka. The dance has many variations, the most popular of which, the war dance, was performed by the All-Blacks rugby team. The robot programmed to perform the ritual dance was a means to build a digital repository of the Maori tradition so that Haka can be performed when humans aren't able to. This was also an attempt to create preservers of traditions at risk of disappearing (Sandoval et al., 2016). However, to truly achieve the purpose of the dance, the robot needs to not just imitate the physical actions but also feel the group sentiment, call onto the meaning generated from shared histories, and perform it with the same mentality as the group mates. This flaw circles back to the anthropocentrism in robotics and AI with the main aim of creating a human-like robot that closely resembles and acts like humans. This may not be in the best interest of social robots learning traditional dances such as rituals that require interaction with their environment at a sensory level. This is a great insight into what is lacking in robotics culture and helps us reimagine the term embodiment in AI.

Humans use dance as a form of storytelling that can be based on any number of creatures from different species. Limiting AI only to imitate human behaviour represents just one of the limitations of anthropocentric AI. Instead, we must re-adjust our expectations to understand AI as a different and separate form of intelligence and accommodate AI-generated experiences as unique experiences on their own (Cross and Ramsey, 2021). Rather than viewing the role of AI and robots in dance as an awkwardly fitting part of the dance culture, perhaps it is time to acknowledge it as a sub-culture emerging from the resistance of their acceptance into the mainstream. Darda et al. (2024) modified the lens used to view AI and robots in research concerning art creation by labelling the animated figures of the robot (robot avatars) and human (human avatars) on screen as "performing agents" and allowing them a sense of autonomy. They attributed choreographies that were supposedly generated using AI algorithms to dances performed by robot avatars and choreographies that were supposedly created by humans to human avatar's performances. While the robot avatar met people's expectations of a robot, the human avatar to a degree remained animated and distinct from human appearances. The aesthetic evaluation for robot avatars was higher compared to human avatars, regardless of the source of choreography (whether it was human-generated, or AI-generated). The study states that lower evaluation could be due to the disappointment that comes with viewing human avatars that did not look very human but robot avatars that were predictably or satisfactorily robotic.

The findings from the above-mentioned study provide a valuable perspective on AI and robots, and their role in dance culture. Viewers have a specific reference for what is evaluated as an aesthetic dance performance. Our expectations of a dance performance are predominantly directed by their human renditions, therefore, using this matrix to evaluate AI robotic rendition would make it an unfair comparison. Having artificially intelligent robots represent AI and robots instead of humans in dance allows viewers to develop a separate aesthetic evaluation for these dances.

References

"Dance Central". Harmonix Music Systems, Inc. 2010. accessed October 18, 2024. https://www.dancecentral.com/

"DanceDanceRevolution A". *Konami Digital Entertainment.* 2016. Accessed October 18, 2024. https://www.konami.com/games/asia/en/products/ddr_a/

Backx, Jacco. "Sand Architects: The Nesting Advantages of Sand Circle Construction by the White-Spotted Pufferfish (Torquigener albomaculosus)." PhD diss., 2023.

Bannerman, Henrietta. "Is Dance a Language? Movement, Meaning and Communication." *Dance Research* 32, no. 1 (2014): 65–80.

Baradaran, Amir. "Towards a Decolonial I in AI: Mapping the Pervasive Effects of Artificial Intelligence on the Art Ecosystem." *AI & Society* 39, no. 1 (2024): 7–19.

Bellaiche, Lucas, Rohin Shahi, Martin Harry Turpin, Anya Ragnhildstveit, Shawn Sprockett, Nathaniel Barr, Alexander Christensen, and Paul Seli. "Humans Versus AI: Whether and Why We Prefer Human-Created Compared to AI-Created Artwork." *Cognitive Research: Principles and Implications* 8, no. 1 (2023): 42.

Benjamin, Walter. "*The Work of Art in the Age of Mechanical Reproduction, 1936.*" New York (1935).

Blacking, John. "Movement and Meaning: Dance in Social Anthropological Perspective." *Dance Research* 1, no. 1 (1983): 89–99.

Best, David N. "Some Problems in the Aesthetics of Dance." *Journal of Aesthetic Education* 9, no. 3 (1975): 105–111.

Brown, Steven, Michael J. Martinez, and Lawrence M. Parsons. "The Neural Basis of Human Dance." *Cerebral Cortex* 16, no. 8 (2006): 1157–1167.

Cross, Emily S., Antonia F. de C. Hamilton, David JM Kraemer, William M. Kelley, and Scott T. Grafton. "Dissociable Substrates for Body Motion and Physical Experience in the Human Action Observation Network." *European Journal of Neuroscience* 30, no. 7 (2009): 1383–1392.

Cross, Emily S., and Richard Ramsey. "Mind Meets Machine: Towards a Cognitive Science of Human–Machine Interactions." *Trends in Cognitive Sciences* 25, no. 3 (2021): 200–212.

Darda, Kohinoor M., and Emily S. Cross. "The Computer, A Choreographer? Aesthetic Responses to Randomly-Generated Dance Choreography by a Computer." *Heliyon* 9, no. 1 (2023): e12750.

Darda, Kohinoor Monish, Aaron Maiwald, Tanvi Raghuram, and Emily S. Cross. "Dancing Robots: Aesthetic Engagement is Shaped by Stimulus and Knowledge Cues to Human Animacy." *Frontiers in Human Neuroscience* 18 (2024): 1413066.

Davies, Stephen. "Evolution, Aesthetics, and Art: An Overview." In Richard Joyce (Ed.), *The Routledge Handbook of Evolution and Philosophy*, 359–371. 2017. Routledge.

Demmer, Theresa Rahel, Corinna Kühnapfel, Joerg Fingerhut, and Matthew Pelowski. "Does an Emotional Connection to Art Really Require a Human Artist? Emotion and Intentionality Responses to AI-Versus Human-Created Art and Impact on Aesthetic Experience." *Computers in Human Behavior* 148 (2023): 107875.

Farah, Charlotte, Lauriane YM Michel, and Jean-Luc Balligand. "Nitric Oxide Signalling in Cardiovascular Health and Disease." *Nature Reviews Cardiology* 15, no. 5 (2018): 292–316.

Filar-Mierzwa, Katarzyna, Anna Marchewka, Aneta Bac, Aleksandra Kulis, Zbigniew Dąbrowski, and Aneta Teległów. "Effects of Dance Therapy on the Selected Hematological and Rheological Indicators in Older Women." *Clinical Hemorheology and Microcirculation* 66, no. 2 (2017): 157–165.

Fong Yan, Alycia, Stephen Cobley, Cliffton Chan, Evangelos Pappas, Leslie L. Nicholson, Rachel E. Ward, Roslyn E. Murdoch et al. "The Effectiveness of Dance Interventions on Physical Health Outcomes Compared to Other Forms of Physical Activity: A Systematic Review and Meta-Analysis." *Sports Medicine* 48 (2018): 933–951.

Freeman, Walter. *A Neurobiological Role of Music in Social Bonding*. eScholarship, University of California, 1998.

Gazzaniga, Michael S. *Human: The Science Behind What Makes Us Unique*. HarperCollins, 2008.

Georgios, Lykesas. "The Transformation of Traditional Dance from Its First to Its Second Existence: The Effectiveness of Music-Movement Education and Creative Dance in The Preservation of Our Cultural Heritage." *Journal of Education and Training Studies* 6, no. 1 (2018): 104–112.

Ho, Rainbow T. H., Ted C. T. Fong, Wai Chi Chan, Joseph S. K. Kwan, Patrick K. C. Chiu, Joshua C. Y. Yau, and Linda C. W. Lam. "Psychophysiological Effects of Dance Movement Therapy and Physical Exercise on Older Adults with Mild Dementia: A Randomized Controlled Trial." *The Journals of Gerontology: Series B* 75, no. 3 (2020): 560–570.

Jeong, Young-Ja, Sung-Chan Hong, Myeong Soo Lee, Min-Cheol Park, Yong-Kyu Kim, and Chae-Moon Suh. "Dance Movement Therapy Improves Emotional Responses and Modulates Neurohormones in Adolescents with Mild Depression." *International Journal of Neuroscience* 115, no. 12 (2005): 1711–1720.

Kosuge, Kazuhiro, Tomohiro Hayashi, Yasuhisa Hirata, and Ryosuke Tobiyama. "Dance Partner Robot-MS Dancer." In *Proceedings 2003 IEEE/RSJ International Conference on Intelligent Robots and Systems (IROS 2003)(Cat. No. 03CH37453)*, vol. 4, pp. 3459–3464. IEEE, 2003.

Knoblich, Günther. "Bodily and Motor Contributions to Action Perception." In *Embodiment, Ego-Space, and Action*, 61–94. Psychology Press, 2008.

Lee, Jounghwa, Han J. Lee, Bommna Ko, and Boni Boswell. "Effects of Creative Dance on Students' Self-Expression and Perceptions of Dance." *International Journal of Applied Sports Sciences* 25, no. 1 (2013): 68–77.

"Living Archive by Wayne McGregor by Wayne McGregor in Collaboration With Artist in Residence at Google Arts & Culture Lab: Bastien Girschig - Experiments With Google." 2019 Accessed October 18, 2024. https://experiments.withgoogle.com/living-archive-wayne-mcgregor

Mingon, McArthur, and John Sutton. "Why Robots Can't Haka: Skilled Performance and Embodied Knowledge in the Māori Haka." *Synthese* 199, no. 1 (2021): 4337–4365.

Mitsunaga, Noriaki, Takahiro Miyashita, Hiroshi Ishiguro, Kiyoshi Kogure, and Norihiro Hagita. "Robovie-IV: A Communication Robot Interacting with People Daily in an Office." In *2006 IEEE/RSJ International Conference on Intelligent Robots and Systems*, 5066–5072. IEEE, 2006.

Mori, Masahiro. "The Uncanny Valley: The Original Essay by Masahiro Mori." *IEEE Spectrum* 6, no. 1 (1970): 6.

Nakaoka, Shin'ichiro, Fumio Kanehiro, Kanako Miura, Mitsuharu Morisawa, Kiyoshi Fujiwara, Kenji Kaneko, Shuuji Kajita, and Hirohisa Hirukawa. "Creating Facial Motions of Cybernetic Human HRP-4C." In *2009 9th IEEE-RAS International Conference on Humanoid Robots*, 561–567. IEEE, 2009.

Or, Jimmy. "Towards the Development of Emotional Dancing Humanoid Robots." *International Journal of Social Robotics* 1 (2009): 367–382.

Pan, Yunhe. "Heading Toward Artificial Intelligence 2.0." *Engineering* 2, no. 4 (2016): 409–413.

Paré, Zaven. "The art of Being Together with Robots: A Conversation with Professor Hiroshi Ishiguro." *International Journal of Social Robotics* 7 (2015): 129–136.

Peng, Hua, Changle Zhou, Huosheng Hu, Fei Chao, and Jing Li. "Robotic Dance in Social Robotics—A Taxonomy." *IEEE Transactions on Human-Machine Systems* 45, no. 3 (2015): 281–293.

Peng, Hua, Jing Li, Huosheng Hu, Keli Hu, Liping Zhao, and Chao Tang. "Automatic Aesthetics Assessment of Robotic Dance Motions." *Robotics and Autonomous Systems* 155 (2022): 104160.

Prinz, Jesse. "Emotion and Aesthetic Value." In *American Philosophical Association Pacific Meeting*, vol. 15. Kluwer, 2007.

Romic, Bojana. "Negotiating Anthropomorphism in the Ai-Da Robot." *International Journal of Social Robotics* 14, no. 10 (2022): 2083–2093.

Šabanović, Selma. "Inventing Japan's 'Robotics Culture': The Repeated Assembly of Science, Technology, and Culture in Social Robotics." *Social Studies of Science* 44, no. 3 (2014): 342–367.

Sandoval, Eduardo Benítez, Omprakash Rudhru, and Qi Min Ser. "The Birth of a New Discipline: Robotology. A First Robotologist Study over a Robot Maori Haka." In *2016 11th* ACM/IEEE *International* Conference on Human-Robot Interaction, 2016

Skjoldager-Nielsen, Daria, and Kim Skjoldager-Nielsen. "Today's Cake Is a Log: Remediating the Intermediality of Hotel Pro Forma's Works in an Exhibition." *The Power of the In-Between* 1, no. 2 (2012): 149.

Suchman, Lucille Alice. *Human-Machine Reconfigurations: Plans and Situated Actions*. Cambridge University Press, 2007.

Vukadinović, Maja, and Slobodan Marković. "Aesthetic Experience of Dance Performances." *Psihologija* 45, no. 1 (2012): 23–41.

Willis, Lauren E. "Why Not Privacy by Default?" *Berkeley Technology Law Journal* 29 (2014): 61.

Wingenroth, Lauren. "How Are Dance Artists Using AI—and What Could the Technology Mean for the Industry?" *Dance Magazine*, July 24, 2023.

6

KEIICHIRO SHIBUYA'S ANDROID OPERA

The Theatrics of Exoticism

Yuji Sone

Founding the music label ATAK in 2002, Japanese composer Keiichiro Shibuya has produced cutting-edge electronic music in various forms, such as piano solos, film scores, and sound installations.[1] Shibuya is one of the most celebrated musicians and composers of his generation, prolifically developing multiform works. In 2012, Shibuya staged *The End*, a concert piece at Yamaguchi Center for Arts and Media (YCAM), that utilised Yamaha's singing generation software Vocaloid, calling it a "Vocaloid opera" (Shibuya 2012). Using the phrase "opera" in a broad and inclusive manner, Shibuya created *The End* as a visual and musical performance without human singers or an orchestra. *The End* featured virtual diva Hatsune Miku, a popular character that is associated with the first product of the Vocaloid series. Hatsune Miku's "concerts" have been staged using large cinematic projection with video mapping technologies. *The End* toured European cities and performed at notable venues in the following year, including a concert at Théâtre du Châtelet, a major opera house and theatre in Paris. At the time of the Paris concert, Shibuya started envisioning an operatic work featuring androids (Billboard Japan 2018). Collaborating with Hiroshi Ishiguro, a famed Japanese roboticist, and Takashi Ikegami, a notable Japanese researcher of artificial life, Shibuya has been producing music performance works with androids since 2015.

This chapter discusses Shibuya's recent experimental operatic works with androids Skelton and Alter 2, 3, and 4: *Scary Beauty* (2017–2018) and *Mirror* (2022–2023). While Skelton and Alter (all versions) were constructed by the team of engineers at Ishiguro's laboratory, Ikegami and his laboratory team developed "neural" systems for these machines

DOI: 10.4324/9781003475972-7

that would allow automated action and reaction. Skelton is of human height with a human-like white face and hands, and wears a smock-like garment, while the metallic components and frames of its body and head remain visible. Alter 2 has similar features. However, Alter 2 consists of a human-sized upper body on a large metal extension rod that is anchored on its base, which allows it to turn to face the musicians on the stage and also turn to face the audience, as well as move the body of Alter 2 vertically up and down. It looks like an enormous, fantastical, therianthropic monster that has a menacing presence like the centaur, a hybrid, mythological creature that is half-human and half-horse: in the case of Alter 2, it is half-human and half-crane machine. Alter 3 is equipped with extra functionality that allows it to recite ChatGPT-generated texts from its mouth and to see with cameras located in both eyes. Alter 4 wears a tiara with metal edges, and it has a more advanced capacity for facial expression. In *Scary Beauty*, which Shibuya produced in a studio setting, and in *Mirror,* in the larger proscenium theatrical settings, Shibuya has posed the android at the centre of the works' staging as a reciter, singer, and conductor of the live orchestra.

Shibuya's android opera utilises technological theatrics that seem to stir ambivalent feelings emerging from dark themes, such as the end of humanity and machinic otherness. His approach and the themes of his android opera have been appreciated at international art and cultural festivals, including those that celebrate culturally eclectic artistic expressions: *Scary Beauty* was featured in the "Meeting Points" programme at the 2017 OzAsia Festival in Adelaide and in Sharjah, UAE, as part of the 2020 exhibition "Inter-Resonance: Inter-Organics, Japanese Performance and Sound Art"; and *Mirror* was included in the Japan Day event programme in the Dubai Expo. The Paris production of *Mirror* in 2023 was included in a "Fête de la musique" (World Music Day) event. Examining video recordings, his comments in interviews, and review articles, I will discuss the above works of Shibuya's android opera in relation to the international contexts in which he presented them, highlighting his strategies for navigating the operatic traditions of the West while at the same time enhancing his stature in Japan.

The Opera and Technology

Opera is a complex and multifaceted artform with a long history that draws from multiple disciplines. First, I will consider the context for Keiichiro Shibuya's use of opera for his android works and his participation in Western operatic traditions, particularly the opera's use of machinic objects and devices in its stagecraft as well as onstage.

European opera of the Baroque era (the 17th and first half of the 18th century), which often explored Greek and Roman mythological subject matter, frequently featured "supernatural interventions, sudden apparitions, and magical transformations" that were presented with "fantastical plots and wondrous machines," according to historian and theorist of opera Gundula Kreuzer (2018, 8). Machinery was "an artistic miracle in and of itself" for Baroque opera (Kreuzer 2018, 8). It is reported, for example, that an operatic production of *La Dafne* in 1608 featured Apollo, the god of sun and light, fighting Python, a fire-breathing dragon with wings that was built as a mechanical costume (Steadman 2021, 142). By promoting his work with the phrase "android opera," Shibuya locates the android as artistically wondrous technology *in and of* opera, as well as the stated narrative. Shibuya's opera is thus also about the developing nature of the android technology in the context of works that he posits as aligned with operatic tradition. The android device in effect frames and shapes his opera. Shibuya takes his technologies as a core aspect of the operatic spectacle, in essence not dissimilar to Baroque opera.

Importantly, Shibuya's android opera exploits ambivalent feelings concerning technology, a theme that emerged from some European operas that embraced humanistic ideals in the Enlightenment period in the late 17th and 18th centuries. For example, the 17th-century French essayist Charles de Saint-Évremond critiqued the use of machines in Italian opera, arguing: "Machines may satisfy the curiosity of ingenious Men, who love Mathematical Inventions, but they'll hardly please persons of good judgment in the Theatre" (Quoted in Kreuzer 2018, 9). According to philosopher of music and aesthetics Lydia Goehr, the 17th-century English opera elevated the opera's aesthetic dimension while concealing the spectacle of its "supporting machinery" (2014, 100). For Kreuzer, critical views of the utilisation of machines in opera reflect an increasing intellectual trend towards the Enlightenment concern for freedom, mastery, and progress, in which the Cartesian mechanistic worldview, where "machines increasingly became codified as objects of scientific study and human progress," dominates (2018, 9). The general demotion of mechanical stage design that featured in the opera was furthered by a surge of comic opera (*opera buffa*) in the 18th century. With its emphasis on sung drama and farcical plots, comic opera required few spectacular machines. In the time of Romanticism, in the late 18th to the mid-19th century, composers began controlling and developing a vision of operatic works as "immersive musical multimedia" that integrate varied artforms (Kreuzer 2018, 12). German composer Richard Wagner's vision of the *Gesamtkunstwerk*, or "total work of art," around the mid-19th century sought to present an influential argument for a unification and interrelationship of various artforms in operatic performances and beyond (Follett and Finger 2011). Stage technologies were intended to be integrated seamlessly within Wagnerian opera.

I argue that Shibuya's android opera presents a unique contemporary case that updates and complicates discussions on the relationship between opera and technology in the traditional format of the proscenium theatre. Some recent operatic works have used robots in traditional opera formats, such as those by Shibuya's contemporary Western counterparts, like Tod Machover's *Death and the Powers* (2010), which uses robotic machinery, and Gob Squad's *My Square Lady* (2015), which features a small humanoid. Significantly, Shibuya's vision for the opera regards an android performer as having substantial agency, equal to that of human performers, questioning the central position of the human in the Western tradition of opera. Since he is a (non-Western) Japanese artist, Shibuya's android opera contains additional layers of meaning: it *works with* the tension between humans and technology in the opera. I argue that Shibuya's operatic works with androids stimulate stereotypical (Western) views of Japan that conflate exoticism and technophilia.

The Opera and the Exotic

It is widely known that Japan was seen as an exotic country in the Far East in the 19th century and the early 20th century, a rich source for the imaginations of Gilbert and Sullivan in their opera *The Mikado or, The Town of Titipu* (1885) and for Giacomo Puccini's *Madama Butterfly* (1904). Satirising the discriminatory injustices of the feudal ruling class of the imaginary town of Titipu in Japan, *The Mikado* is a comic opera that depicts a love triangle with "queer and quaint characterizations" that frame and reference the Japanese as the "yellow peril," a formulation of the era (Lee 2010, xix). In *Madama Butterfly*, the character Cio-Cio-San (Madama Butterfly), "a self-sacrificing Japanese Heroine" who performs a tragic suicide at the end, has been discussed as an exemplary case of the Orientalism of the day, reflecting "the gendered dynamics of East-West relations founded upon unequal power relations" (Yoshihara 2004, 975). The use of "yellowface" treatment of the Japanese characters, that is, performed by white Caucasian performers, in *The Mikado* and *Madama Butterfly* has been criticised (Lee 2010, xv; Hara and Yoshihara 2024, 143). Nevertheless, the longevity of these productions, still staged in the present, demonstrates the ongoing fascination in mainstream Western opera for fetishised and often kitsch images of Japan and Japanese people by opera fans in the West (Lee 2010, xvi). I will return to this later, in terms of French Japonism in the 21st century.

Importantly, Shibuya's android opera also involves his own Japanese nationality as a key element. Just like Igor Stravinsky, deploying his Russian background in the 20th century, Shibuya locates himself in "the 'exotic' category" (Sheppard 2014, 802) in the 21st century. Shibuya is conscious of being a Japanese composer who creates operatic works that participate in a longstanding Western tradition. It would certainly have been a notable event for a non-Western opera composer to stage a work at Théâtre du Châtelet in Paris.

For centuries, the operatic form reflected essentially Western cultural and class capital as "a spectacle of well-dressed affluence and social exclusivity" (Kotnik 2016, 1) and a symbol of "high culture" (Kotnik 2016, 3). Paris, along with London, was regarded as one of "the capital cities that had come to define cosmopolitan taste and social practices." Hosting operatic works contributed to this view (Weber 2007, 160). Shibuya explains that a non-Western composer like himself cannot compete with Westerner composers if he just follows the established styles and systems of Western music; the use of Japanese traditional musical instruments alone would simply be regarded as ornamental in this overdetermined context (Murota 2022). Therefore, Shibuya realises that, in opera, he must avoid a simplistic use of Japanese cultural language or traditional Japanese music (New National Theatre Tokyo 2021). Instead, Shibuya's strategy is to position non-human performers such as Hatsune Miku or an android in the centre of an operatic work, visibly challenging the humanistic tradition, and, importantly, utilising stereotypes to create a mix of attraction and repulsion for the Western opera audience.

Android "Otherness" and the Cultural Otherness of Japan

While Shibuya's approach correlates to perspectives in more recent opera studies that challenge the central position of human agency and subjectivity (Kreuzer 2018; Morris 2019), his android opera should be recognised as a cultural-critical response from his position as a Japanese composer. At the 2017 OzAsia Festival in Adelaide, a major art festival in Australia with a focus on Asia, Shibuya presented the Vocaloid opera *The End* at Dunstan Playhouse. He also premiered *Scary Beauty*, a 35-minute chamber opera that featured the singing android Skelton at Space Theatre with the Australian Art Orchestra. Peter Knight, the artistic director of the Australian Art Orchestra, programmed *Scary Beauty* under the Meeting Points programme, which "brings together vast Asian and Australian music styles" at the festival to stimulate cultural dialogue and the development of new ideas (Adelaide Festival Centre 2017, 14–17).

On one level, the context of the OZAsia Festival already framed *Scary Beauty* as a work of an Asian Other, the Japanese Other. It is therefore inevitable that some features of Skelton would be regarded as "Japanese." Reviewer Murray Bramwell describes Skelton's porcelain-like, neutral white mask with "dark darting eyes" as "Japanese features," as it may have resembled a Noh mask if one reads into its static expression (Bramwell 2017). Similarly, the unnatural movements of Skelton reminded reviewer Peter Wilkins of "Butoh" (Wilkins 2017), a form of avant-garde dance originating in late 1950s and early 1960s Japan. In *Scary Beauty*, the android Skelton was dressed in a burgundy smock, a garment prepared by renowned Japanese design house Sacai, and it was wearing high-laced Doctor Martens boots. It is therefore not surprising that some reviewers regarded Skelton as "female" (Bramwell 2017; Lenny 2017) even though it was meant to be "asexual" (ATAK Tokyo 2017). The reviewers' assumptions concerning the android as "female" may have been based on a familiar Japanese character in the opera context: that is, *Madam Butterfly*, or the Western view of "Asia" as "the feminine."

The art context that emphasised cultural exchange might have catalysed a resurfacing of the longstanding stereotypical image of Japan as a technophilic country, as in the phrase used by Frederik L. Schodt to describe the view of "Japan as a Robot Kingdom" in the 1980s in the English-language literature (1988, 15). A long-held view of Japan as technocentric has relatedly been transformed to "techno-Orientalism," which regards Japan as "the original techno-Orient" (Roh, Huang, and Niu 2015, 3) and "a problem for the West" (Morley and Robins 1995, 147). Shibuya's provocative comments on the decentring of the *human* position in his opera could also have contributed to a renewed operatic Orientalism.

The Demotion of Human Performers

The title of the opera *Scary Beauty* directly reflects Shibuya's positive inclination towards the figure of the android. The oxymoronic title aims to engage positively with two opposing feelings, such as awe and fear, trust and mistrust, and original and fake, giving equal weight to both. This paradoxical expression reflects Shibuya's investment in the idea of a sustained duality, a characteristic of much Japanese cultural and traditional thinking, that, he states, suggests that everything contains two sides of things, such as good and bad, or safety and danger, and that he wants to engage with such paradoxes positively in his operatic works (Ando 2018a).

Highlighting the non-humanness of the android's voice and movements in *Scary Beauty*, Shibuya states that he wants audiences to appreciate "beauty" in its strangeness, and regard "coexistence" with such mechanical beings with "a new feeling [that lies] between scary and beauty" (ATAK Tokyo 2017). By foregrounding the android Skelton, Shibuya explores contrasts between humans and the android, and associated contradictory emotions toward the figure of the android – positive and negative feeling, pleasantness and unpleasantness, trust and mistrust, or awe and fear. At the same time, Shibuya's intention is to highlight grey areas between the two poles. Shibuya's positive vision for simultaneous accommodation of opposites can be alienating with regard to the Western tradition of the either/or dualism.

Shibuya rejects the traditional formal style of the opera, in which performers stand at the centre of the production narratively and visually. In *Scary Beauty*, while Shibuya was conducting and playing the piano, Skelton "sang" excerpts of selected texts not originally intended to be set to music by well-known authors such as Michel Houellebecq, Yukio Mishima, William S Burroughs, and Brion Gysin, who collaborated with Burroughs, all authors and artists who put forward a bleak view of the human condition. Their texts were shown in subtitles projected on rear screens, which also showed live footage of the android and pre-recorded images. Shibuya selected their texts that portray images of the world "after all mankind became extinct" (ATAK Tokyo 2017). The android functioned as a surrogate ventriloquist's doll for Shibuya, generating "alienating" effects (Murota 2022).

With the more advanced Alter 2 robot, Shibuya revised and represented an expanded version of *Scary Beauty* in 2018, further exploring his ideas concerning the android as the central focus of his opera. Adding excerpts from Ludwig Wittgenstein's work, the new version of *Scary Beauty* as a 50-min chamber opera piece located the Alter android at the centre of the work as the conductor of a 28-piece orchestra, along with Shibuya at the piano and a larger video projection that showed a live image of the performing android. This version of the opera was presented at Miraikan (the National Museum of Emerging Science and Innovation) in Tokyo as part of the ALIFE 2018 conference (the International Conference of Artificial Life Research). Alter 2 was equipped with an artificial "neural" network system that allowed it to generate movements as responding to external stimuli, such as "the sound of the orchestra or its own voice," according to artificial life researcher Takashi Ikegami, a core collaborator of Shibuya's since 2000 who has developed the computer programmes for Shibuya's android operas (Billboard Japan 2018). With the sophisticated functionality developed by Ikegami, Alter 2's central role as the conductor was facilitated by preprogrammed inputs concerning the "lyrics, melody and number of bars" structuring the music (Billboard Japan 2018).

Shibuya has stated that he is interested in reversing the power relationship between human musicians and the android, according to journalist

Kenji Ando (2018b). Shibuya wants to explore humans' possible creative reactions when led by the robot because he thinks that humans have become an "obstacle" for creativity and would need technological enhancement in order to advance (Ando 2018a). Shibuya explains that his intention for *Scary Beauty* with Alter 2 was to test new creative possibilities that might open up by locating an android as a conductor, thereby taking the dominant position in the operatic work (Ando 2018a). The orchestra musicians needed to adapt to the android's unusual movements and non-human logic. As well, Alter 2 can sing at superhuman speed. Alter 2's non-human voicing of the famous writers' very human words could potentially bring about a very different "reality" and new "emotions" for the audience of *Scary Beauty* (Ando 2018a). Shibuya's aim here is to give rise to new meanings.

For Shibuya to fully realise his vision for Alter's centrality in his opera work, the android platform needed to be developed further. Alter 2 was not designed for music performance, and it was malfunctioning after the Tokyo show in 2018 (Fujikawa 2021). With support from the Japanese IT company Mixi Inc and Warner Music Japan, Alter 3, an upgraded android with a better actuator, a camera for each eye, and a speaker system in its mouth, was developed. For Shibuya, these new features would enable the android's performance to have "more seamless" expressions and movements in its semi-autonomous conducting and singing performance (quoted in Billboard Japan 2018). Ikegami explained that his ultimate research goal concerning the Alter android was to understand "how to construct an actual mind in the artificial brain and body" that could conduct an orchestra (Billboard Japan 2018). Such comments could affirm the persisting stereotypical image of Japan as the technophilic cultural Other, which would stimulate interest in non-Japanese overseas audiences. Indeed, Shibuya presented *Scary Beauty* with Alter 3 in *Hi, Robot! Mensch Maschine Festival*, an art festival with the theme of technology and art in Düsseldorf, Germany, in 2019, and in Sharjah, UAE in 2020 as part of the exhibition "Inter-Resonance: Inter-Organics, Japanese Performance and Sound Art" at the Sharjah Art Foundation.

Coexistence of the Old and the New

Taking advantage of increasing overseas interest, Shibuya's next project *Mirror* (2022) exploited another stereotypical image in the West of Japan as a hybrid culture of the "old" and the "new."

At Expo 2020 Dubai, a World Expo that was delayed to March 2022 due to the COVID-19 restrictions in 2020, the half-hour android opera *Mirror*, featuring Alter 3, premiered. The android was displayed in the context of internationally recognisable signs of traditional Japanese Buddhism. Earlier, for the Japan Day event in the 2020 Dubai Expo "Connecting Minds, Creating the Future," Shibuya was asked to use signs of Japanese traditional culture in his android opera (The Hollywood Reporter Japan 2023). For that event, Shibuya contacted Eizen Fujiwara, a Buddhist monk and performer of Nanzanryū Shōmyō, a style of recital chanting with a history of over 1,200 years in Japan, who had been performing publicly with Shibuya in works including *Heavy Requiem* (2019), a collaborative performance involving Shibuya's electronic music and Fujiwara's chanting at Ars Electronica, Linz, Austria (Shibuya 2019). In *Mirror*, a group of Buddhist monks, including Fujiwara's fellow monks, who were wearing colourful traditional prayer robes, participated as a chorus. Shibuya also conducted a 45-member symphony of primarily white Western concert musicians from UAE's NSO Symphony Orchestra, along with Alter 3, for an audience of mixed nationalities (Expo 2020-2022). Alter 3 was situated downstage while Shibuya played the piano and conducted the orchestra at stage right, next to the android. The monks were positioned at stage left next to Alter 3. The orchestra members were upstage, just behind Shibuya, Alter 3, and the monks, in front of a large video screen that projected video images of the performers designed by video artist Justine Emard, which were mixed with close-up vision of the real-time facial expressions of Alter 3. The work presented a deliberate mix of cultural ingredients in which the Alter 3 android was the focus. Reviewer Shigeru Ogawa describes this performance of *Mirror* as "woven together by unique 'outsiders'" in the Arabian Peninsula context (2022).

The key features of *Mirror* were Alter 3's speech and singing performance in contrast to the monks' chanting. *Mirror* started with Alter 3's recitation, in its amplified, echoing "voice," of Shibuya's statement: "Android is a mirror. Music is a mirror. It is a reflection of yourselves," while the orchestra provided background music (Shibuya 2022). Eizen Fujiwara, the leader of the Buddhist group, joined Alter 3 with his chanting. The android appeared to be "responding" to Fujiwara's chanting with facial expressions and subtle finger and hand movements. While it was "listening" to the chanting, Alter 3 continued to recite some philosophical questions by Shibuya: "What is the boundary between existence and non-existence? What is the boundary between past and future? And where do they exist?" (Shibuya 2022) When Shibuya started playing the piano, Alter 3 repeated the statement: "Let's celebrate this new experience together" at the end of the first section (Shibuya 2022). Alter 3 continued performing its recitation and singing with the monks' chanting, the orchestral music, Shibuya's piano playing, and electronic sound, including making a statement in the midst of the performance that the lyrics for the third and the fourth sections were written by ChatGPT (Expo 2020-2022). When Alter 3 sang in response to Fujiwara's chanting, Fujiwara, in turn, chanted back with his palms together. He later explained that "[he] faced the android and chanted with [his] whole heart, intending to become one and merge with it" (Ogawa 2022). In essence, Fujiwara describes, with utter sincerity, performing a Buddhist practice to identify with Alter 3. Fujiwara's Buddhist approach to engaging with Alter 3 on stage may well have presented a fascinating contrast between the "old" and the "new" Japan for many of the non-Japanese audience members at the Jubilee Stage at the Dubai EXPO.

Shibuya wasted no time promoting the success of *Mirror* in Dubai. Later in 2022, Shibuya negotiated a three-day performance run of *Mirror* at the Théâtre du Châtelet in Paris, proposing that it would premier in 2023 on June 21st, which, for more than 40 years, has been a date for a notable annual event to mark "Fête de la musique" (World Music Day) (The Hollywood Reporter Japan 2023).

In the concluding section of this chapter, I discuss the Paris performance of *Mirror* as a cumulative product of Shibuya's operatic works, relating to the key aspects of his operas that I have discussed so far: the presentation of Japan's cultural otherness, as it is perceived in the West; the centring of an android and related demotion of human performers; and the juxtaposition of signs of traditional and technological Japan. At an event that illuminated the strategic focus of Shibuya's android opera, he offerred a self-Orientalising intended for consumption overseas *and* for success in Japan, precisely because of its praiseworthiness in the international context.

In Conclusion: *Mirror* at Théatre du Châtelet and 21st-Century French Japonism

Shibuya returned to the Paris venue in 2023 with the new version of *Mirror* extended to 70 minutes and featuring the more sophisticated Alter 4 (Figures 6.1 and 6.2). This updated android was equipped with an artificial tongue and more facial articulations under its white mask that allowed the android to display richer expressiveness, and with an increased number of joints on its body (from 43 to 53) to enable more dynamic movements (Osaka University of Arts 2023).

As it was connected to ChatGPT, during the performance, Alter 4 spontaneously responded to the monks' chanting. The 2023 version of *Mirror* also incorporated *Dai-hannya Tendoku*, a "rolling reading" of the *Great Perfection of Wisdom Sutra*, in which monks chant a folded roll of the Sutra by opening it in the air and moving pages between their hands. This act is normally performed to abbreviate a reading of the sutra, which can

FIGURE 6.1 Android Opera MIRROR (2023). © Théatre du Châtelet, Photo by Thomas Amouroux.

FIGURE 6.2 Android Opera MIRROR (2023). © ATAK, Photo by Claude Gassian.

take several days if one tries to read it properly. Developing from the Dubai version, the Paris production of *Mirror* also included the Buddhist performance of *Gyōdō*, which is a walk around a Buddhist statue or a Buddhist shrine while chanting. The five participating monks conducted a *Gyōdō* around Alter 4 as if they regarded the android as an embodiment of the Buddha and were paying their respects to it. With its superhuman reading capability, Alter 4 could read the sutra and respond immediately, while humans would require days. These acts presented the typical image of Japan as a paradoxical mix of old and new, an interweaving of advanced technology and unchanged traditions. Overall, *Mirror* represented a techno-exotic image, thanks to the traditional elements that were incorporated in the work, seamlessly presented alongside AI technologies, of a contemporary and even future-ready Japan for the 21st-century French audience. Alter 4, a giant monster with the form of a human with technologically advanced attributes, performed an exaggerated hybridity of Japanese culture, a fused form that may have been simultaneously alienating and attractive.

It is possible to speculate that *Japonismes 2018: les âmes en resonance*, a large-scale cultural project held in 2018–2019 in Paris to showcase Japanese culture and art, set up the right context for Shibuya's return to the Théâtre du Châtelet in 2023 with an opera featuring an android performing with Buddhist monks. *Japonismes 2018* was produced by the Japanese government under the agreement between the late Prime Minister Shinzo Abe and French President François Hollande at the Japan-France Summit Meeting in 2016 to mark the 160th anniversary of Japan-France diplomatic relations in 2018 (The Japan Foundation 2018).

Abe discussed the term "Japonism" in his speech at the opening of *Japonismes*: "Introduced to France under the term 'Japonism,' Japanese culture stimulated the creativity of French artists in the nineteenth century" (Abe quoted in Kang 2019, 23). Unlike Edward Said's famous concept "Orientalism," which asserts the centrality of the West as early as ancient Greece, positing the East or the Orient as the disparaged Other (Said 1991), in Abe's use of Japonism, the otherness of Japanese culture and art is a positive factor for "stimulating new cultural creativity," leading to "the spiritual resonance of Japan and France" (Abe quoted in Kang 2019, 24). However, from a 21st-century perspective on French Japonism, the android singer Alter 4 can be regarded as a contemporary version of an exotic Oriental character in opera, perhaps equivalent to Cio-Cio-San in Puccini's *Madama Butterfly*, but without the negative connotations.

Crucially, the exotic markers of Japan in *Mirror* are to do with Shibuya's *own* exoticising and participation in a traditional "operatic orientalism [that] includes both the experience of attraction and revulsion in encountering the exotic" (Sheppard 2014, 811). I want to suggest that Shibuya's opera productions in Paris were, paradoxically, a product of his "self-Occidentalisation," immersing himself in the Western operatic tradition, while his use of Buddhist chanting was a strategic choice to engender "self-Orientalisation." Alter 4 could also be understood as the emblem of Shibuya's strategic, doubled gesture of "self-Occidentalisation" and a concurrent "self-Orientalisation." Alter 4, as a product of advanced technology, could be said to emblematise the Japanese history of imitating modern, Western technologies since the 19th century. For Shibuya, staging a contrast between the new and the old, and between Japan and the West, which appears as an exotic duality, was his way of competing and demonstrating his unique contributions in the Western opera context (Murota 2022). Alter 4 functioned as a binding sign for Shibuya's android opera's strategic entanglement with this tradition on his own terms.

Shibuya's strategy of self-Occidentalisation and self-Orientalisation relates to a similar approach adopted by the previous generations of Japanese musicians. For example, Haruomi Hosono is known as a key member of the Japanese electronic rock band Yellow Magic Orchestra from the late 1970s who deployed a strategic ambiguity of cultural traits in popular music-making. For musicologist Shuhei Hosokawa, such an approach reflects the view that the construction of post-war Japanese identity was based on simultaneous "self-occidentalisation and self-orientalisation of Japanese culture" (Hosokawa 1999, 114).

I am arguing that Shibuya strategises similarly for his android opera to conjure up an appetite for what might be called "Japonism" by non-Japanese audiences, which, in turn, would elevate recognition of his work in Japan and overseas. Unlike Hosono, Shibuya is not cynical nor satirical but instead tries to balance "the Western" and "the Eastern" impulses and approaches, even as he may realise that he is only feeding the Western appetite for an imagined, exotic Japan. I suggest that learning from the Buddhist concept that everything is connected with everything else in the universe, Shibuya wants to explore resonances between oppositions such as the East and the West, humans and non-humans, and male and female, rather than insisting on the differences between them (The Hollywood Reporter Japan 2023). This idea of harmonious correlation differs from the idea of unity as a totality, as in the Wagnerian *gesamstkunstwerk*, or total artwork, which sought "a new form of unity or synthesis" for a "desired but impossible totality" (Follett and Finger 2011, 8). The *Mirror* production in Paris reflects Shibuya's aspiration for an achievable state that is not a unity, but a state of harmonious coexistence of "unresolvable" elements that do not in fact need to be "resolved" in order to be appreciated.

Whether trading in Japonism or not, the Paris season of *Mirror* would very likely have been interpreted by the Japanese public as proving that

Shibuya's opera was accepted by an "authentic" opera audience in Paris. In real terms, *Mirror*'s success would also attract music and art producers overseas. Indeed, Shibuya states that, immediately after the Paris show, he was approached by the Lisbon Film Festival to present his work at the festival's closing ceremony (Fujikawa 2024). *Android Aria – Seeds of Prophecy* (2023), which consisted of Shibuya's piano playing, his electronic music, and Alter 4's singing, though without a symphony orchestra, was presented in November 2023. Due to its limited budget, the show was presented as a concert featuring Shibuya and Alter 4.

The vision for and practices of Keiichiro Shibuya's android opera are a 21st-century extension of the operatic tradition's incorporation of new technologies, and Shibuya's android opera reflects opera's history of exoticism and "other-ing" for Shibuya's own purposes, both aesthetic and commercial. The android Alter has become, at least for the time being, the unique signature and hallmark of Shibuya's brand, "android opera," providing him with opportunities to continue his work in both Japan and overseas.

Note

1 I am representing Japanese names in the English-language way: that is, given name first, followed by family name. Long vowel sounds are indicated by diacritical marks, unless in common usage in Romanised form (e.g., Tokyo, not Tōkyō).

References

Adelaide Festival Centre. 2017. "Ozasia Festival 2017 Brochure." Issuu. July 25, 2017. https://issuu.com/festivalcentreadelaide/docs/ozasia_brochure_issuu_2017_version_

Ando, Kenji. 2018a. "Andoroido wo tōshite ningen wa jibun jishin wo miru (Humans Seeing the Self through the Android)." *The Huffington Post Japan*. August 26, 2018. https://www.huffingtonpost.jp/entry/scary-beauty-002_jp_5c5aae17e4b0cd19aa9465bc

Ando, Kenji. 2018b. "Robotto to ningenn wa koi ni ochiiruka (Can a Robot Fall in Love with a Human?)." *The Huffington Post Japan*. August 25, 2018. https://www.huffingtonpost.jp/entry/scary-beauty-001_jp_5c5aae14e4b0cd19aa946572

ATAK Tokyo. 2017. "Past performance - Adelaide." *ScaryBeauty*. September 30, 2017. http://scarybeauty.com/archive/archive170930-en.html

Billboard Japan. 2018. "Japan's 'Scary Beauty' Android Opera Is a Metaphor for Our Relationship with Tech." February 7, 2018. https://www.billboard.com/music/music-news/japan-android-opera-scary-beauty-interview-8463632/

Bramwell, Murray. 2017. "Meeting Points: Scary Beauty and The End: Vocaloid Opera (OzAsia, Adelaide)." *Daily Review*. October 7, 2017. https://dailyreview.com.au/meeting-points-scary-beauty-keiichiro-shibuya-australian-art-orchestra-skeleton-space-theatre-adelaide-festival-centre/

Expo 2020-2022. "Japan: Android Opera, MIRROR." Expo2020 Dubai. 2022. http://virtualexpo.world/

Follett, Danielle, and Anke Finger. 2011. "Dynamiting the Gesamtkunstwerk: An Introduction to the Total Artwork." In *The Aesthetics of the Total Artwork*, edited by Anke Finger and Danielle Follett, 1–25. Baltimore: Johns Hopkins University Press. 10.56021/9780801895821

Fujikawa, Takahiro. 2021. "Inside the Mind of Keiichiro Shibuya Part 3 The Innovativeness of a Singing Android and Super Angels." TOKION. February 17, 2021. https://tokion.jp/en/2021/02/17/inside-the-mind-of-keiichiro-shibuya-part3/

Fujikawa, Takahiro. 2024. ""Massive Life Flow; Inside the Mind of Keiichiro Shibuya" Part 13 - TOKION." TOKION. February 29, 2024. https://tokion.jp/en/2024/02/29/massive-life-flow-13/

Goehr, Lydia. 2014. "The Concept of Opera." In *The Oxford Handbook of Opera*, edited by Helen M. Greenwald, 0. Oxford University Press. 10.1093/oxfordhb/9780195335538.013.005

Hara, Kunio, and Mari Yoshihara. 2024. "Unpinning Madama Butterfly." *American Quarterly* 76 (1): 141–59.

Hosokawa, Shuhei. 1999. "Soy Sauce Music: Haruomi Hosono and Japanese Self-Orientalism." In *Exoticism in Post-War Popular Musi*, edited by Philip Hayward, 114–44. New Barnet, UK: John Libbey Publishing.

Kang, Tae-woong. 2019. "The Ukiyoe Boom and Twenty-First Century Japonism." *Seoul Journal of Japanese Studies* 5 (1): 23–43.

Kotnik, Vlado. 2016. *Opera as Anthropology: Anthropologists in Lyrical Settings*. Newcastle upon Tyne: Cambridge Scholars Publishing.

Kreuzer, Gundula. 2018. *Curtain, Gong, Steam*. Oakland, CA: University of California Press.

Lee, Josephine. 2010. *The Japan of Pure Invention: Gilbert and Sullivan's The Mikado*. NED-New edition. Minneapolis: University of Minnesota Press.

Lenny, Barry. 2017. "Review: OzAsia Festival 2017: Scary Beauty at Space Theatre, Adelaide Festival Centre." Broadway World. October 2, 2017. https://www.broadwayworld.com/japan/article/BWW-Review-OZASIA-FESTIVAL-2017-SCARY-BEAUTY-at-Space-Theatre-Adelaide-Festival-Centre-20171002

Morley, David, and Kevin Robins. 1995. *Space of Identity: Global Media, Electronic Landscapes and Cultural Boundaries*. London and New York: Routledge.

Morris, Christopher. 2019. "Casting Metal: Opera Studies after Humanism." *The Opera Quarterly* 35 (1–2): 77–95. 10.1093/oq/kbz015

Murota, Naoko. 2022. "Keiichiro Shibuya: What's New about the Two Android Operas, the One at the New National Theater and the Other at Dubai Expo?" ONTOMO. January 22, 2022. https://ontomo-mag.com/article/interview/keiichiro-shibuya01/

New National Theatre Tokyo, dir. 2021. *Super Angels - Special Talk Session Vol.1*. https://www.youtube.com/watch?v=6ipXitNgLhQ

Ogawa, Shigeru. 2022. "The World Premiere of Keiichiro Shibuya's Android Opera, 'MIRROR,' at Expo 2020 Dubai." TOKION. May 8, 2022. https://tokion.jp/en/2022/05/08/massive-life-flow-5/

Osaka University of Arts. 2023. "Keiichiro Shibuya's Android Opera "MIRROR" will be performed in June at the Théâtre du Châtelet in Paris, starring Alter 4 from the Android Lab, which will excite the audience in Paris." PR TIMES. April 18, 2023. https://prtimes.jp/main/html/rd/p/000000078.000044215.html

Roh, David S., Betsy Huang, and Greta A. Niu. 2015. "Technologizing Orientalism: An Introduction." In *Techno-Orientalism: Imagining Asia in Speculative Fiction, History, and Media*, edited by David S. Roh, Betsy Huang, and Greta A. Niu, 1–19. Asian American Studies Today. Brunswick, New Jersey: Rutgers University Press.

Said, Edward W. 1991. *Orientalism: Western Conceptions of The Orient*. London: Penguin Books.

Schodt, Frederik L. 1988. *Inside the Robot Kingdom: Japan, Mechatronics, and the Coming Robotopia*. Tokyo and New York: Kodansha International.

Sheppard, W. Anthony. 2014. "Exoticism." In *The Oxford Handbook of Opera*, edited by Helen M. Greenwald, 795–816. Oxford University Press. 10.1093/oxfordhb/9780195335538.013.036

Shibuya, Keiichiro. 2012. "Vocaloid Opera — THE END | Théatre Du Châtelet, Paris." ATAK. 2012. http://atak.jp/en/theater/the_end/

Shibuya, Keiichiro. 2019. "Heavy Requiem – Buddhist Chant 'Shomyo' + Electronics." ATAK. 2019. http://atak.jp/en/theater/heavy-requiem/

Shibuya, Keiichiro. 2022. "Android Opera — MIRROR." ATAK. 2022. http://atak.jp/en/theater/android-opera-mirror-dubai-expo/

Steadman, Philip. 2021. *Renaissance Fun*. London, England: UCL Press.

The Hollywood Reporter Japan. 2023. "Keiichiro Shibuya, Interview before The Android Opera 'MIRROR' at The Théâtre du Châtelet in France." The Hollywood Reporter. June 17, 2023. https://hollywoodreporter.jp/music/24878/

The Japan Foundation. 2018. "Japonismes 2018." 2018. https://www.jpf.go.jp/e/about/area/japonismes/

Weber, William. 2007. "Opera and the Cultural Authority of the Capital City." In *Opera and Society in Italy and France from Monteverdi to Bourdieu*, 160–80. Cambridge University Press. 10.1017/CBO9780511481734.011

Wilkins, Peter. 2017. "Scary Beauty - Meeting Points -OzAsia Festival." Canberra Critics Circle. October 2, 2017. https://ccc-canberracriticscircle.blogspot.com/2017/10/scary-beauty-meeting-points-ozasia.html

Yoshihara, Mari. 2004. "The Flight of the Japanese Butterfly: Orientalism, Nationalism, and Performances of Japanese Womanhood." *American Quarterly* 56 (4): 975–1001.

PART TWO

Performing Arts Cultures and Robots/AI Developments

7

CHIKAMATSU, MORI, AND THE UNCANNY VALLEY

Karl F. MacDorman

Introduction

Chikamatsu Monzaemon[1] (1653–1724) is Japan's most celebrated playwright. Dubbed the profession's guardian deity, he wrote at least 40 plays for kabuki and 114 for *ningyō jōruri*[2] (hereafter, *jōruri*), a form of musical puppet theatre he transformed and popularised (Kawatake 1988, Figure 7.1). Chikamatsu's only known work of theatre criticism is reported in an interview with a close friend and *jōruri* devotee, the Confucian scholar Hozumi Ikan (1692–1769). The interview presents Chikamatsu's theory of realism, published posthumously, in the preface to Hozumi's (1738/2016) *Naniwa Miyage* (Kan 2019; Pastreich 2011).

Chikamatsu's insights can inform contemporary discourse on character and interaction design in robotics and AI, especially as it relates to the *uncanny valley*. The uncanny valley graphs how affinity for robots, dolls, and puppets plunges into eeriness when they resemble humans too closely. The concept, first proposed in 1970 by the Japanese robotics professor Mori Masahiro (2012), has taken hold in the West this century. Chikamatsu's theory explains how to balance realism to circumvent the uncanny valley and maintain emotional engagement. The goals of today's designers resonate with Chikamatsu's ambition to bring soulless creations to life, which is why his advice remains relevant.

In subsequent sections, I recount Chikamatsu's biography, translate his interview with Hozumi, reinterpret his theory of realism, and apply it to robotics and AI research. As a translator of Mori's essay and a researcher publishing extensively on the uncanny valley, I explore new connections

DOI: 10.4324/9781003475972-9

FIGURE 7.1 Rō Ikkan in Chikamatsu Monzaemon's *The Battles of Coxinga*, puppeteered by Yoshida Bunshō III at the National Theatre in Tokyo, February 13–28, 1982. (Courtesy of the C.V. Starr East Asian Library, Columbia University).

between Chikamatsu's theory, human psychology, and contemporary robotics. My aim is to show the universality of Chikamatsu's insights, implicit in Mori's uncanny valley concept, in developing social robots.

The Playwright's Life and Contribution

Chikamatsu was born Sugimori Nobumori, the second son of a samurai who served as a medical doctor under daimyo (lord) Matsudaira Tadamasa (Miyoshi 2016). After losing his office, Chikamatsu's father moved to Kyoto. Chikamatsu briefly served as a page to the nobleman Ichijō Ekan, then entered *Gonshō-ji* monastery in Ōmi province, where he studied classical Japanese literature, music, court ceremony, etiquette, and Buddhism (Iwao et al. 1975).

Chikamatsu started writing *jōruri* plays at 25 or 26 and continued for the rest of his life. In 1684, he began a 30-year collaboration with the narrator Takemoto Gidayū (1651–1714). Takemoto opened his theatre in Osaka with *The Soga Heir*, which Chikamatsu had written for Uji

Kaganojō's theatre in Kyoto the year before. Beginning with Takemoto's production of *Kagekiyo Victorious* in 1685, set after the Genpei War (1180–1185), Chikamatsu innovated a new style of *jōruri* with more realistic, expressive, and riveting plots (Yoshikawa 1965). In 1706, when the kabuki actor Sakata Tōjūrō (1647–1709) retired, Chikamatsu moved to Osaka to dedicate himself to writing *jōruri* for Gidayū's theatre (Kawatake 1988).

The bulk of Chikamatsu's *jōruri* repertoire consists of 90 *jidaimono* (plays depicting historical events and figures). These period pieces include romances and epic battles involving nobles and samurai. Masterpieces of this genre include *The Snow Woman* (1705), *The Battles of Coxinga* (1715), and *The Soga Brothers at Kaikei Mountain* (1718). The remainder consists of 24 *sewamono* (plays depicting the contemporary life of ordinary people). These domestic dramas have received the most enduring acclaim (Chikamatsu 1961). Set in the present, focused on family conflicts, sometimes featuring an antihero and exposing life on society's fringes, they marked a transgressive departure from earlier *jōruri* (Gerstle 1996). A character inspired by an orphan, miscreant, or murderer could leap from the news headlines to the stage. Chikamatsu's domestic dramas include *The Love Suicides at Sonezaki* (1703), *The Love Suicides at Amijima* (1720), and *The Woman-Killer and the Hell of Oil* (1721). In Woman-Killer, the oil merchant's son, Yohei, falls heavily into debt and turns to blackmail and murder. Chikamatsu's portrayal of Yohei as an antihero rather than a villain resonates with contemporary audiences.

Chikamatsu elevated puppet theatre to the apex of dramatic art. His plays, the first to incorporate townspeople as a literary genre, along with similar stories by poet and novelist Ihara Saikaku (1642–1693), continue to shape Japanese literature. His domestic dramas, known for their acute insight into family and romantic relationships, explore the tension between desire and duty (Iwao et al. 1975). That this tension nearly always resolves in tragedy indicts the rigid feudal order, unable to fulfil individual needs in an urbanising society.

My translation of Hozumi's interview with Chikamatsu follows.

Preface to Naniwa Miyage by Hozumi Ikan

Some years ago, when I visited Chikamatsu, this is what he told me:

1 *Jōruri* differs from other forms of storytelling. As the puppets' role is paramount, every verse must spring to life through movement. Moreover, *jōruri* vies for patrons with nearby kabuki theatres and the artistry of their live actors. To move the audience, the playwright must evoke

various emotions using soulless wooden dolls. Without extraordinary effort, it is hard to write a play of great renown.

Luckily, while reading *The Tale of Genji*[3] in my youth, I stumbled upon this passage: "At the time of a seasonal court feast, snow had fallen heavily, blanketing a *tachibana* tree.[4] A guard was dispatched, but no sooner had he brushed off the *tachibana* than the pine tree beside it, still bending under the weight of snow, recoiled its branches with reproach." With these strokes of a writing brush, the author breathed life into a soulless plant. That is because the pine tree, envying the care the *tachibana* received, shook the snow from its own branches. Does this not create the impression of a being coming to life?

This passage made me realise how to imbue my *jōruri* with soul. Descriptions of scenery,[5] not to mention narration and dialogue, must be steeped in emotion, lest the work fail to stir deeply. This is akin to what poets call *evocative imagery*. For instance, when extolling in poetry the magnificent landscapes of Matsushima or Miyajima,[6] unless one writes with exaltation, it would be like gazing inattentively at a beautiful woman's portrait. That is why each phrase must be rooted in emotion.

2 Whatever the verse, using too many particles like *te* [and], *ni* [to], and *ha* [as for] somehow degrades it. Yet untalented writers, confusing *jōruri* with poetic forms like *waka* and *haikai*,[7] try to fit their verses into a 5–7 syllable pattern, resulting in superfluous particles. For example, if they should write *Toshi mo yukanu musume wo* [a girl not yet come of age], they instead write *Toshiha mo yukanu, musume wo ba* [as for a girl not yet come of age, indeed]. This fixation with counting syllables makes the dialogue sound unrefined. Though short and long lines are arranged in poetry, *jōruri* consists of song with string accompaniment, so the melody determines each line's length. Should a playwright adhere strictly to poetic metre, the verses may prove awkward to chant. As I have not attempted this, my scripts contain few particles.[8]

3 *Jōruri* used to be like today's *saimon* [street storytelling and lyrical ballads]. It lacked substance and flair. After leaving Uji Kaganojō's troupe to write plays for Takemoto Gidayū, I chose my words with great care to lift *jōruri* to a high art. For example, the nobility, samurai, and lower tiers have distinct ranks and social positions, and each must be depicted accordingly in speech and manner. Even within the samurai, the daimyo, *karō* [chief retainer], and lower ranks, whose stipends vary, must be distinguished. This ensures that the audience empathises with each character.

4 *Jōruri* depicts life as it happens. But in the interest of art, it also exposes hidden aspects. It is standard practice in kabuki for an *oyama* [male actor who portrays female roles] to deliver lines that a real woman would not

utter. By talking openly, the character reveals her true feelings and circumstances. If a playwright were instead to model the character on a real woman, thereby concealing her motives and intentions, the performance – in all its accuracy – would fail to delight. Lines unbefitting a woman invite criticism from those who overlook the need for this device, but such departures from realism should be viewed as art. Likewise, when a villain acts too cowardly or a fool too buffoonishly, this should be viewed as art rather than a presentation of reality, and the audience should accept it as such.

5 Writers, declaring melancholy essential to *jōruri*, pepper their plays with expressions like *How sorrowful!*[9] Otherwise, they have the lines chanted in a tear-drenched voice in the *bunya-bushi* style. [Okamoto Bunya, 1633–1694, was an early *jōruri* playwright.] These methods are absent from my work. In my plays, sorrow arises solely from each character's adherence to duty. As the plot thickens, the burdens of duty intensify. The melody and dialogue are so tightly constructed that the audience is deeply moved. To lament "How sorrowful!" distracts from its causes, diminishing the final impression. Sorrow must arise, not on command, but spontaneously.

Furthermore, to admire a landscape like Matsushima in a single breath, "Ah! What a splendid view!" gives the impression of there being nothing left to say, thus diminishing it. To praise the scenery, take instead an indirect approach: point out several of its features to let its appeal be grasped naturally. This advice holds for anything of this kind.

6 Hozumi remarked: I have heard it said that society these days [early 18th-century Japan] will not accept things unless they seem well reasoned, realistic, and convincing. Many aspects of old stories no longer persuade. They deem kabuki actors skilful only if their performances resemble real life. An actor portraying a *karō* must resemble a real *karō*, one portraying a daimyo must resemble a real daimyo, and so on. Audiences will not tolerate, as they had in the past, the kinds of artifice that would only fool a child.

Chikamatsu answered: This argument may sound plausible but shows no grasp of the essence of art. Art abides in the narrow gap between the flesh and the skin, the real and the unreal. With today's preference for realism, an actor portraying a *karō* is expected to mimic the speech and gestures of a real *karō*. But does that mean a daimyo's *karō* should smear his face with powder and paint like the actor? Or, granted that his *karō* wears no makeup, would it then prove as entertaining to watch the actor perform on stage unpainted with a bald head and unkempt beard? This is what I mean by the narrow gap between the flesh and the skin: unreal yet not unreal, real yet not real. Somewhere between the two, amusement unfolds.

This observation relates to the story of a court lady who became enamoured with a man. Their passion was mutual, but the man could not call on her, as she was secluded deep within the imperial palace. She could only catch chance glimpses of him at court by peering through slits in the bamboo blinds. She was so madly in love with him that she ordered a wooden statue to be carved in his likeness. This was no ordinary doll. Its face and body differed not a hair's breadth from the man's. The colouration of his complexion, the pores of his skin, his earholes, nostrils, and mouth, and even the number of teeth inside, down to the tiniest detail, were copied exactly. Indeed, since the man had posed beside the figure while it was being made, it differed from him only in its absence of a soul. Yet when the court lady drew near to see it, though the doll was an exact replica, her enthusiasm instantly cooled. She somehow found it vacant and eerie. Her passion for the man cooled as well. Just having the doll around proved so disturbing that she soon discarded it.

Reflecting on this, if we reproduce exactly a being of flesh and blood, even the ravishing Yang Guifei, at some point, our fondness runs out. [Yang Guifei, 719–756, was an imperial consort and one of the four beauties of ancient China.] Thus, whether painting a figure or carving it in wood, the artist must use broad strokes in places, even while aiming to stay true to its form, as it is a fabrication. Human likeness of this type is, after all, what inspires people's love. Likewise, in composing a plot, the playwright must use broad strokes in places while still following the original story, for this skill ultimately delights the audience. One should also keep this in mind when composing theatrical dialogue and the like.

Chikamatsu's Divisions of Realism

Chikamatsu is often admired in film circles for proposing a theory of realism (Kan 2019). On this view, the scriptwriter should not aim for surface realism – a set free of fancy costumes, wigs, and makeup – but for conceptual realism: The characters should be separated and ordered to show their role and status by depicting their differences through demeanour, gesture, and attire.

Conceptual realism, according to Chikamatsu, can enable the audience to identify with the characters and empathise with them. A play with conceptual realism – combined with expressive imagery and a compelling plot – will evoke the right feeling at the right time (Mangan 2015). Joy and sorrow will arise from what the events unfolding on stage mean for each character, as the character's hopes and desires are aroused and thwarted. This progression of plot-driven emotions holds the audience in suspense and transforms the cast's actions into art. The performance need not be highly realistic, but the audience must experience it as meaningful because meaning makes the performance *feel* real.

Chikamatsu's theory states that art dwells in a narrow gap between the unreal and the real, the inauthentic and the uncanny. However, this metaphor should not be taken literally (Lee 2000). It reflects one powerful style of art. With the benefit of hindsight, we see that other styles do not abide in a narrow gap but in a wide expanse, from the abstract, geometric patterns of Oskar Fischinger's *An Optical Poem* (1937, echoed in Disney's *Fantasia*, 1940) to the gritty realism of Vittorio De Sica's *Ladri di Biciclette* (1948), shot on location with ordinary townsfolk. At the low end of realism, the movements of two triangles and a circle tell a story that exhibits their different personalities and allows viewers to feel empathy for them (Heider and Simmel 1944); at the high end of realism, security camera footage amuses viewers on social media. None of these examples feels fake; none feels eerie. Thus, we can experience art and meaning at any level of surface realism – even when venturing into the uncanny.

Nevertheless, Chikamatsu's theory of realism rings true for *jōruri*, kabuki, and social robotics for two reasons. First, Chikamatsu's aim was not just to create art but *popular* art. To this end, he employed conceptual realism to stir the passions. Unlike *nō*, *jōruri* and kabuki were the popular theatre arts of his day. The stage was ablaze with action and drama, which, for kabuki, sometimes sparked a riot (Mezur 2005; Ortolani 1995). While Chikamatsu's new *jōruri* exhibited artistic merit, evidenced by its critical acclaim, he also multiplied the art form's popularity. Consider, for example, *The Battles of Coxinga*, which depicted Prince Coxinga's resistance against the Manchu conquest. At a time when a play might typically run for a few weeks, it ran for an unprecedented 17 months. Shakespeare, whose plays were also hugely popular in his time, similarly used conceptual realism in characterisation: "Shakespeare does not paint individuals but individualises classes. [He distinguishes in each character] the signs of a class of nature, midway between his general nature and his individual peculiarities" (Whipple 1888, 49). Both Chikamatsu and Shakespeare were popular because they used conceptual realism to create archetypical and relatable characters and holistic and moving stories.

Second, Chikamatsu's theory rings true for *jōruri* and kabuki because conventions of symbolism and stylisation are integral to these art forms and limit their surface realism. To a contemporary audience, these conventions may obscure meaning, especially when combined with situations, social norms, and language made alien by the passage of time. In the U.S., *kabuki theatre* is a metaphor for elaborate but empty political posturing that leaves matters unresolved. Ironically, this is the opposite of how kabuki – or *jōruri* – would have been experienced during the Genroku period (1688–1704) and successive decades. Conflicts on stage were resolved, often tragically, by the action and dialogue, which the audience could follow because the situations, social norms, and language were still contemporary. Conventions clarified a play's meaning. By supporting conceptual realism, *symbolism* and *stylisation* enhanced the play's popular appeal.

Symbolism and stylisation have also been used to this effect in robot performances in Japan and the U.S. (Sone 2017). For example, R2-D2 from *Star Wars* employed distinctly non-human sounds, movements, and visual cues to convey personality and emotion.

The juxtaposition Chikamatsu makes between conceptual and surface realism, both of which he placed between the unreal and the real, may be interpreted as forming a graph (Figure 7.2). As indicated, the old *jōruri* was unrealistic and unconvincing. It lacked authenticity, relying on graceless theatrics. The narrator would *tell* the audience what to think and feel about a landscape instead of allowing the audience to draw its own conclusion.

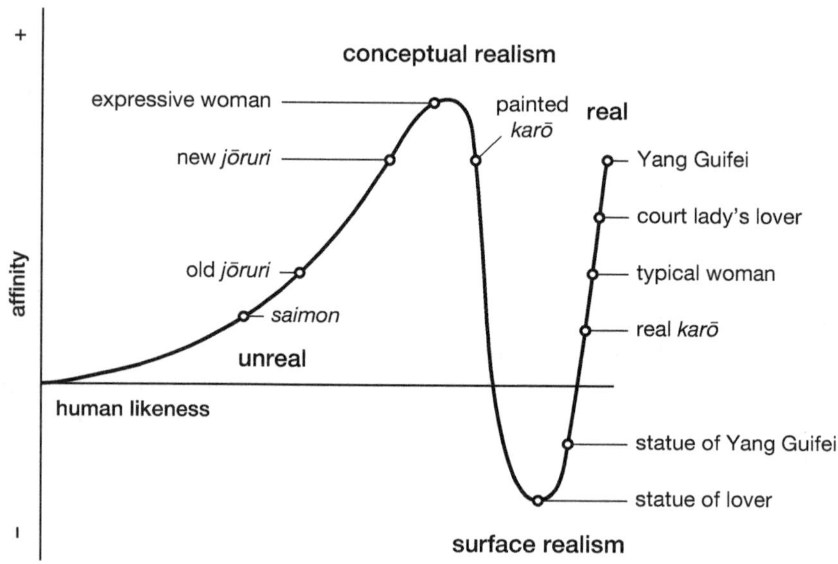

FIGURE 7.2 Chikamatsu's theatre criticism implies that conceptual realism increases affinity through characterisation and plot-driven emotions. Surface realism employs exact reproductions of reality, which are eerie. The audience will often prefer conceptual realism because its departure from reality can heighten a play's artistic merit. (Credit: MacDorman).

Then, from the story of the envious pine in *The Tale of Genji*, Chikamatsu learnt how to enliven his puppets with emotion. Movement, dialogue, and evocative imagery would *show* events unfolding. The drama would hold the audience under its spell, free of the jarring proclamations of an overwrought narrator. In essence, the dramatic technique "show, don't tell," attributed to Anton Chekhov (1860–1904), had been proposed by Chikamatsu two centuries before it was popularised in the West (Lubbock 1921). This technique was essential to his new *jōruri*.

Chikamatsu also contrasts real life with kabuki's conceptual realism: During the Edo period, a real woman was expected to keep her feelings buried, while a female character could express them openly, thus advancing the plot. Like a Shakespearean soliloquy, though unrealistic, this openness allows the audience to understand the character's state of mind. As is common in drama, the audience is expected to suspend disbelief. Similarly, though a real *karō* presents his face without makeup, a kabuki actor could use makeup to portray the *karō's* disposition and state of mind. For example, the colours red and blue represent good and evil, respectively; lines on the face, hands, and forearms represent blood vessels throbbing with anger or passion; makeup shows whether the character is bold or reticent, thoughtful or reckless; and clothing indicates the character's social status. Though one might think the stylised kabuki versions of the woman and *karō* were false, Chikamatsu affirms that such departures from reality are art. Thus, a woman and *karō*'s role are placed near the first peak in the graph.

Chikamatsu then sets up another contrast – this time between surface realism and real life: The court lady is unnerved by the statue of her lover, rendered with uncanny accuracy. Even a statue of Yang Guifei, whose beauty was legendary, would quell the viewer's affection if reproduced precisely. With these parables, Chikamatsu reaffirms that art sometimes demands departures from reality.

Conceptual Realism in Mori's Uncanny Valley

In today's Japan, *jōruri* and kabuki have been crowded out by other forms of entertainment. Puppet theatre is a dying art in the West as well.

Chikamatsu, however, is making broader observations about visual storytelling and the aesthetics of anthropomorphism. These observations apply to new forms of media like computer animation and virtual and augmented reality. Given Chikamatsu's goal – to create emotional engagement by breathing life into soulless puppets – his observations on *jōruri* apply especially well to social and humanoid robots. Their aesthetics were addressed in a 1970 essay by Mori Masahiro (2012), then a professor at the Tokyo Institute of Technology. Chikamatsu's theory of realism foreshadows Mori's uncanny valley concept, aligning with it in thought-provoking ways (Poulton 2014).

Mori's essay describes how our affinity for robots varies with their human likeness. Industrial robots, designed only to perform specific functions, do not resemble the factory workers they replace, and most people feel little affinity for them (Figure 7.3). By contrast, toy robots, which have a head, torso, and limbs, delight children. Humanoid robots, which imitate human interaction using sophisticated sensors and actuators, can be even more engaging. Thus, with each human feature added, should we not expect our affinity for robots to increase?

Presumably, if a robot could be built that appeared – in every aspect – identical to a healthy, active person, it would engender even greater affinity. But Mori observes that this goal is perilous. As the robot approaches but fails to become indistinguishable from humans, our response abruptly turns from affinity to revulsion. Mori presents two thought experiments. In the first, a woman reaches out to shake what looks like a human hand but is, in fact, a myoelectric hand. The discrepancy arouses uncertainty about whether a living being is animate. In the second, a man rummages around lifeless mannequins that suddenly awaken. This aberration sparks uncertainty about whether non-living objects are inanimate. Jentsch

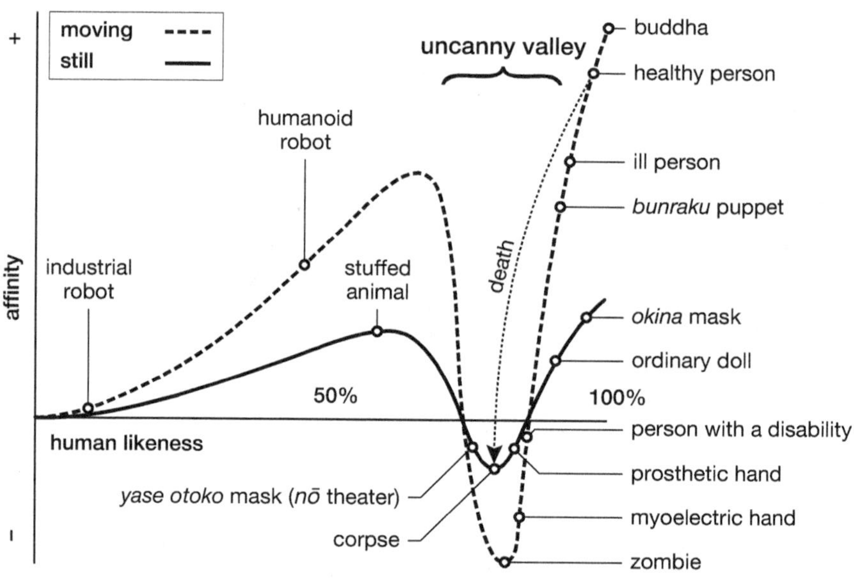

FIGURE 7.3 Mori's uncanny valley plots the relation between an entity's human likeness and the affinity it evokes. Movement steepens the slopes. In 2005, he added *buddha*. (Credit: MacDorman).

(1906/1997) identifies uncertainty of either kind as eliciting an uncanny sensation.[10] Mori attributes this to the "uncanny valley phenomenon," the title of his seminal essay.

To interpret Mori's essay, Chikamatsu's distinction between conceptual and surface realism is invaluable. We find a blending of these concepts in Mori's placement of items along his graph's human likeness dimension. For example, the *okina* (old man) mask from *nō* theatre is highly stylised – with a broad nose, closed eyes, and evenly spaced wrinkles, symbolising wisdom, longevity, and prosperity. In its *surface* realism, it does not appear more human than the *yase otoko* (emaciated man) mask. Nevertheless, Mori's graph presents the *okina* mask as almost human, much more so than the *yase otoko* mask (Figure 7.3). The placement of the masks only makes sense when one considers their degree of *conceptual* realism: The *yase otoko* mask represents a ghost from hell, which is indeed less human and more eerie than an old man.

In 2005, Mori indicated that a statue of the Bodhisattva Maitreya, considered our world's next buddha, could be placed above and to the right of the healthy person in his graph (Mori et al. 2012). In placing the statue at the pinnacle of human likeness, Mori prioritises the *concept* the statue embodies – buddha as the *ideal* human – over its physical appearance. Again, his graph reflects conceptual realism more than surface realism.

A similar blending of conceptual and surface realism occurs on the curve for moving entities. Mori (2012) observes that on close inspection, *bunraku* puppets appear less real than prosthetic hands. However, when animated on stage, their performance in its aesthetic mastery appears so human as to captivate the audience. Thus, his graph plots a *bunraku* puppet as more human than the prosthetic hand – or even its limbless user. As Chikamatsu surmised, the puppets are engaged in art, which is more beguiling than reality. I depict this interpretation in Figure 7.2 by making the first peak higher than the second. The audience feels an affinity for the puppets, not just because their movements appear human but because each puppet embodies an individual identity (LePage 2021). The puppet achieves this by portraying its character and the character's relationships in a meaningful story (MacDorman and Cowley 2006).

How Narrative Empathy Falls Apart

I have proposed that an artificial entity's surface realism disrupts emotional empathy for characters and, thus, the enjoyment of visual narratives (MacDorman 2019). As Chikamatsu observed from the court lady's encounter with her lover's statue, surface realism can turn affinity into eeriness. Although affinity tends to increase with an artificial entity's human likeness and perceived sentience, when a human-looking entity violates human norms, this elicits the uncanny valley effect (MacDorman 2024).

Zillmann and Cantor's (1977) affective disposition theory predicts that a viewer's emotional responses to a character are shaped by moral judgements about the character's actions. If the character's actions are deemed benevolent or justifiable, the viewer will come to regard the character favourably, and the viewer's emotional responses will align with the character's emotions (Raney 2004). Conversely, if the actions are deemed malevolent, the viewer will come to regard the character unfavourably, and the viewer's emotional responses will misalign with the character's emotions.

I proposed empathy disruption theory, which predicts that heroes will be most vulnerable to the uncanny valley's effects (MacDorman 2019). In addition to traditional heroes, we can feel an affinity for antiheroes, even a rogue like Shakespeare's Falstaff or a murderer like Chikamatsu's Yohei, both of whom fell victim to circumstances (Gerstle 1996; Raney 2004). Because context determines our affinity for a character, it also determines the character's vulnerability to the uncanny valley and, in turn, our feelings about the performance. The impact of context can be generalised beyond the dramatic arts to any relatable character, whether in computer animation, virtual and augmented reality, social and humanoid robotics, or real life.

In sum, affinity depends on many factors apart from human likeness: Whether the entity embodies an individual identity, enacts a story, or qualifies as a hero, antihero, or villain, the particulars of the story, whether it makes sense, has artistic merit, etc.

Applying Chikamatsu's Theory to Social Agents

To test empathy disruption theory, we set up an online interactive scenario where participants role-played as a patient interacting with a doctor (MacDorman 2019). The theory predicts that an artificial entity's surface realism can disrupt emotional empathy, particularly for heroes, by eliciting the uncanny valley effect, thus reducing narrative enjoyment. The experiment has a 2 × 2 × 2 between-groups design where the doctor is either a hero or villain, the narrative has either a happy or sad ending (the doctor is awarded a fellowship or sued for malpractice), and the doctor and his office are either depicted using computer animation or a real actor. The computer-animated doctor and office were digital doubles of the real ones.

Many effects were as predicted: Participants had greater emotional empathy for the hero than the villain. They also had greater sympathy, cognitive perspective-taking, and narrative enjoyment. Participants preferred just endings (hero rewarded, villain punished) to unjust endings (hero punished, villain rewarded).

Some effects were counter to prediction. Participants enjoyed the computer-animated scenario more than the real one, even though the computer-animated doctor rated eerier. Eeriness ratings were influenced by the story, resulting in the villain being rated eerier than the hero. Nevertheless, narrative enjoyment operated somewhat differently for the tragic hero, depending more on the participant's transportation to the virtual world than emotional empathy for the doctor. Eeriness may disrupt narrative empathy more in a different genre, like romance, or using a different medium, like a physical robot.

Lessons for Social Robotics

Chikamatsu's principles for *jōruri* can be adapted to the experience design of social robots:

Anthropomorphism's value to *jōruri* was illustrated in the passage about the envious pine. In social robotics, Paro, modelled after a baby harp seal, leverages its zoomorphism to engage users (Sone 2017). The robot turns its head towards sources of sound, feigning attention. This trick enables Paro to tap into psychological mechanisms that cause us to attribute human intentions to this mindless robot (Turkle 2007).

Empathy for a character is elicited in *jōruri* and social robotics by aligning the character's actions to the viewer's values (MacDorman 2019).

Individuation of character types and roles in *jōruri* translates into social robots that exhibit context awareness when interacting with different types of people (e.g., customer vs. employer or adult vs. child). They express their identity, purpose, and function through their overall performance (LePage 2021; Sone 2017).

"*Show, don't tell*," according to Chikamatsu, indicates meaning should arise through the story, not its narration. Thus, a social robot should reveal its inner state through its deeds and relationships, not just its statements (MacDorman and Cowley 2006).

Simplicity is essential to creating an interaction flow that feels organic. In *jōruri*, that involves the words following the melody, and in social robotics, it could involve the melody following the emotional prosody of speech (Savery, Zahray, and Weinberg 2021).

Symbolism in kabuki is exemplified by coloured grease paint, indicating a character's nature, status, and emotions. Similarly, the original Sony Aibo, a robotic pet dog, uses coloured LEDs, tail wags, and body movements. Kabuki and *nō*'s symbolism and stylisation appear in robot performances in Japan; they encourage audiences to suspend disbelief and perceive the robots as magical, more than machines (Sone 2017).

The *uncanny valley* warns contemporary designers not to aim for perfect human likeness but instead for stylised, emotionally engaging designs (Mori 2012).

Though three centuries have passed since Chikamatsu shared his criticism of *jōruri* and kabuki with his friend Hozumi, his views can inform how we design robots and other technologies that present themselves as social agents. The goal, to breathe life into a soulless contrivance, remains the same. As does the path forward, to design robots that enact compelling, conceptually realistic stories, whether these robots appear human or not.

Notes

1 The family-name-before-given-name convention in Japanese is followed throughout.
2 Macrons indicate long vowel sounds.
3 *The Tale of Genji*, the world's first novel, describes life at the Heian court (Murasaki 2008).
4 The blossoms of this citrus tree are fragrant, and the fruit resembles mandarin oranges.
5 This refers to *michiyuki*, lyrical descriptions of scenery during a journey.
6 Matsushima Bay's pine-covered islands and Miyajima Island's shrine and floating *torii* gate are among the three most admired views in Japan.
7 *Waka* and *haikai* are classical and comic Japanese poetry.
8 Despite this claim, Chikamatsu often followed the 5–7 pattern.
9 In classical literature *aware* denotes any moving experience (Motoori 2013).
10 Mori conceived the uncanny valley independently of Jentsch (1997) and Freud (2003; Jochum and Goldberg 2015).

References

Chikamatsu, Monzaemon. *Major Plays of Chikamatsu*. Translated by Donald Keene. New York: Columbia University Press, 1961.

Freud, Sigmund. *The Uncanny*. Translated by David McLintock. New York: Penguin, 2003.

Gerstle, C. Andrew. "Hero as Murderer in Chikamatsu." *Monumenta Nipponica* 51, no. 3 (1996): 317–356. 10.2307/2385613

Heider, Fritz, and Marianne Simmel. "An Experimental Study of Apparent Behavior." *The American Journal of Psychology* 57, no. 2 (1944): 243–259. 10.2307/1416950

Hozumi, Ikan. *Naniwa miyage hottan – Chikamatsu monzaemon no kyojitsu himaku-ron* [*Preface to Memories of Naniwa: Chikamatsu Monzaemon's Theory of Realism in Art*]. Tokyo: Hōbundo, 2016.

Iwao, Seiichi, Tarō Sakamato, Keigo Hōgetsu, Itsuji Yoshikawa, Tadashi Kobayashi, and Shizue Kanazawa. "47. Chikamatsu Monzaemon (1653–1724)." In *Dictionnaire historique du Japon* 3: 33–35. Tokyo: Maison Franco-Japonaise, 1975.

Jentsch, Ernst. "On the Psychology of the Uncanny." Translated by Roy Sellars. *Angelaki* 2, no. 1 (1997): 7–16. 10.1080/09697259708571910

Jochum, Elizabeth, and Ken Goldberg. "Cultivating the Uncanny: *The Telegarden* and Other Oddities." In *Robots and Art: Exploring an Unlikely Symbiosis*, edited by Damith Herath, Christian Kroos, and Stelarc, 149–175. Singapore: Springer, 2015.

Kan, Kyōko. *Chikamatsu jidai jōruri no sekai* [The World of Jōruri during the Time of Chikamatsu]. Tokyo: Perikansha, 2019.

Kawatake, Shigetoshi. *Chikamatsu Monzaemon.* Tokyo: Yoshikawa Kōbunkan, 1988.

Lee, Sang-Kyong. "Edward Gordon Craig and Japanese Theatre." *Asian Theatre Journal* 17, no. 2 (2000): 215–235. 10.1353/atj.2000.0018

LePage, Louise. "The Importance of Realism, Character, and Genre: How Theatre Can Support the Creation of Likeable Sociable Robots." *International Journal of Social Robotics* 13 (2021): 1427–1441. 10.1007/s12369-020-00637-w

Lubbock, Percy. *The Craft of Fiction.* London: Jonathan Cape, 1921.

MacDorman, Karl F. "In the Uncanny Valley, Transportation Predicts Narrative Enjoyment More Than Empathy, But Only for the Tragic Hero." *Computers in Human Behavior* 94 (2019): 140–153. 10.1016/j.chb.2019.01.011

MacDorman, Karl F. "Does Mind Perception Explain the Uncanny Valley Effect? A Meta-Regression Analysis and (De)Humanization Experiment." *Computers in Human Behavior: Artificial Humans* 2, no. 1 (2024): 1–14. 10.1016/j.chbah.2024.100065

MacDorman, Karl F., and Stephen J. Cowley. "Long-Term Relationships as a Benchmark for Robot Personhood." In *Proceedings of the 15th IEEE International Symposium on Robot and Human Interactive Communication*, 378–383, September 6–9, 2006, University of Hertfordshire, Hatfield, UK.

Mangan, Bruce B. "The Uncanny Valley as Fringe Experience." *Interaction Studies*, 16, no. 2 (2015): 193–199. 10.1075/is.16.2.05man

Mezur, Katherine. *Beautiful Boys/Outlaw Bodies: Devising Kabuki Female-Likeness.* New York: Palgrave Macmillan, 2005.

Miyoshi, Shūichirō. "*Chikamatsu Monzaemon no tsūsetsu ni aratana*" [*New Developments in the Prevailing View on Chikamatsu Monzaemon*]. *Fukui Shinbun*, October 17, 2016.

Mori, Masahiro. "The Uncanny Valley." Translated by Karl F. MacDorman and Norri Kageki. *IEEE Robotics and Automation* 19, no. 2 (2012): 98–100. 10.1109/MRA.2012.2192811

Mori, Masahiro, Zaven Paré, Chihiro Minato, and Emmanuel Grimaud. "Le Bouddha dans le robot: Rencontre avec le professeur Masahiro Mori." *Gradhiva* 15 (2012): 162–181. 10.4000/gradhiva.2393

Motoori, Norinaga. *Genji monogatari: Tama no ogushi—Mono no aware-ron* [*The Tale of Genji: A Little Jewelled Comb—A Theory on Poetics*]. Tokyo: Tama Tsūshin-sha, 2013.

Murasaki, Shikibu. *Genji Monogatari* [*The Tale of Genji*]. Translated by Yosano Akiko. *Suetsumuhana* Chapter 6. The Safflower. Tokyo: Kadokawa Bunko, 2008.

Ortolani, Benito. *The Japanese Theatre: From Shamanistic Ritual to Contemporary Pluralism.* Princeton, New Jersey: Princeton University Press, 1995.

Pastreich, Emanuel. *The Observable Mundane: Vernacular Chinese and the Emergence of a Literary Discourse on Popular Narrative in Edo Japan*. Seoul, Korea: Seoul National University Press, 2011.

Poulton, Cody. "From Puppet to Robot: Technology and the Human in Japanese Theatre." In *The Routledge Companion to Puppetry and Material Performance*, edited by Dassia N. Posner, Claudia Orenstein, and John Bell, 280–293. New York: Routledge, 2014.

Raney, Arthur A. "Expanding Disposition Theory: Reconsidering Character Liking, Moral Evaluations, and Enjoyment." *Communication Theory* 14, no. 4 (2004): 348–369. 10.1111/j.1468-2885.2004.tb00319.x

Savery, Richard, Lisa Zahray, and Gil Weinberg. "Emotional Musical Prosody for the Enhancement of Trust: Audio Design for Robotic Arm Communication." *Paladyn, Journal of Behavioral Robotics* 12, no. 1 (2021): 454–467. 10.1515/pjbr-2021-0033

Sone, Yuji. *Japanese Robot Culture: Performance, Imagination, and Modernity*. New York: Palgrave Macmillan, 2017.

Turkle, Sherry. "Authenticity in the Age of Digital Companions." *Interaction Studies* 8, no. 3 (2007): 501–517. 10.1075/is.8.3.11tur

Whipple, Edwin P. *The Literature of the Age of Elizabeth*. Boston and New York: Houghton, Mifflin, and Company, 1888.

Yoshikawa, Eiji. *Nihon ongaku no rekishi* [*History of Japanese Music*]. Osaka: Sogensha, 1965.

Zillmann, Dolf, and Joanne R. Cantor. "Affective Responses to the Emotions of a Protagonist." *Journal of Experimental Social Psychology* 13, no. 2 (1977): 155–165. 10.1016/S0022-1031(77)80008-5

8

TOWARDS EMBODIED AI

Design Approaches for Robots in Opera

Elizabeth Jochum, Tim Hopkins, Chris Kiefer and Evelyn Ficarra

Introduction

Live performance has long been a site for experimenting with new and emerging technologies. This is especially true for opera, where technology has functioned as a catalyst for spectacular scenery and special effects and also a tool for augmenting and extending human performance. From the elaborate flying machines in Corneille's 1650 *Andromede* to the giant robotic set featured in Robert Lepage's 2012 production of *The Ring*, opera's capacious form has proven conducive for developing and testing new technologies. As robot and AI technologies become more sophisticated, robots begin to take centre stage as performers.[1] This chapter describes our ongoing investigations with the *Robot Opera* project to explore how robots and AI might contribute to novel forms of creative expression. We are particularly interested in the topic of embodied AI and how to design machine performers that are responsive and adaptive to their environment. We are motivated by the desire to understand how AI tools might help artists move beyond working with programmable machines towards developing autonomous robot performers that can interact with and learn from performance contexts. Our research has resulted in new tools and creative outputs, including some practical considerations for how multidisciplinary teams can approach AI and machine learning tools in opera.

Opera's operational language requires the convergence of different expressive agents in systems that create physical and emotional spaces for imaginative speculation. Opera unites technologies that create new performative spaces and new instrument designs with live performers, and these

DOI: 10.4324/9781003475972-10

virtuosic interactions require new approaches to musical and scenic composition. Opera rejects a literalist view of realism, announcing its artificiality while simultaneously engaging audiences at a profoundly emotional level through music. These features enable us to imagine AI as an emotional construction as well as a physical one, and to rethink how emotional experiences are constructed through technology within Western performance conventions.

While our initial research focused on composing operas using commercial humanlike robots (Nao; Pepper; Mesmer), we gradually adopted a more bottom-up approach. Instead of imitating human forms and speech, we asked what would happen if we looked beyond the human and instead designed robot performers around the principle of relationality. This pivot prompted us to adopt an exploratory, systems-based approach for modelling behaviours such as listening, mirroring, echoing, and attending to sonic and musical features, as well as physical gestures and behaviours. We found combining analogue approaches such as traditional puppetry and musical improvisation with digital tools in workshops to be a very productive approach.

We recognise our research as part of a rich lineage of media-art works and performances that trace back to the cybernetic artworks that began in the 1960s.[2] Artists have long explored how information systems could function as aesthetic material and developed strategies for generating creative output using computers. New AI and machine learning models continue this investigation by introducing more variation and complexity in outputs and focusing on emergent behaviours. Whereas early computational music followed rules-based approaches for classifying and generating music in a certain musical style, the new class of AI tools (e.g. diffusion models, generative adversarial networks [GANs], and variational autoencoders) allows machines to generate novel musical outputs based on vast training data sets. However, it is one thing to generate creative content through visual or sonic output on a computer, and quite another to render this output with embodied and expressive machines in live performance.

Live performance is by definition an emergent form, regardless of whether computers or algorithms are involved. In opera, composers, musicians, singers, and dancers all interact and draw upon tactile and sensuous knowledge that arises from embodied experiences shaped by physical interactions with instruments, the environment, sound, space, and other bodies. Tim Ingold describes this phenomenon as the *textility of making*, where artists and materials engage in an artful and responsive negotiation of feeling and form. Art works and performances are not fixed, but works in progress that involve emergent processes where artists learn to *follow the materials*. Our understanding of artistic composition resonates with the emergent view of artmaking, where artists do not impose form on matter but bring together diverse materials, "combining or redirecting their flow in the anticipation of what might emerge" (Ingold, 2009, p. 94). The emergent view of art and live performance does not view the material world as passive or subservient to artists, programmers, and composers, but insists on a creative, collaborative, and ongoing negotiation with and through material. Our workshops set out to explore how machine learning tools might facilitate this process with robots by combining computation and embodied interaction with musical expression and composition.

Robots in opera typically appear in one of three forms – dynamic set pieces, humanlike characters, or non-anthropomorphic characters functioning either as props or metaphorical humans that are either preprogrammed or tele-operated by an operator offstage. From a technical standpoint, these robots are more like technological puppets, and not unlike the androids and other humanoid automata that have captured the public imagination for centuries (Schaeffer, 1999; Jochum and Murphey, 2014). The humanoid robots in Seinendan Theatre's *Sayonara* (2010) and Rimini Protokoll's *Uncanny Valley* (2019) meticulously imitate human features and expressions, but even robots that do not aim at verisimilitude still evoke powerful abstractions of the human form, either by reducing it to a kind of parody or else projecting the robot as an extension of the human performer.

When it comes to robot voice, the default choices are either a recorded human voice or a synthetic, "electric" sounding voice, even when the robot's physical appearance is not humanlike. The trope of giving human voices to robots is replicated, albeit with more sonic artistry, in the robot opera repertoire. Wade Marynowski's *Robot Opera* (2015) features non-humanoid robots that are imbued with electronic sounds including, at times, heightened robotic speaking voices. The non-humanoid robots in Tod Machover's *Death and the Powers* (2010) are given human operatic voices as extensions of the human performers who sing live both on and offstage (Torpey, 2012). Another approach can be found in the case of musical robots, where robots communicate through musical expression executed on analogue or digital instruments. One of the more advanced musical robot platforms, Shimon, developed by Gil Weinberg, uses machine learning to compose songs and sing using a synthesised voice (Savery et al., 2021). These developments show how machine learning tools can be applied to robotics, combining multimodal perception to generate rich, expressive, and collaborative interactions between human artists and machines.

Developing machine and human attention models using machine learning tools is a rapidly evolving field. To develop these models, engineers make decisions about which features are salient and optimise models to respond to these features. But these models are usually developed with very applied, domain-specific contexts in mind, and the results are sometimes not explainable even to the developers. Furthermore, these models are rarely developed with theatrical performance in mind. If opera makers are going to make use of data-driven generative music tools, it is important they be designed in such a way that makes intuitive sense for composers and performers. We tasked ourselves to reimagine robot design and embodied AI from a relational standpoint, and to think differently about what new musical interactions between people and robots might look like. Our goal was to leverage a robot's physical affordances so it might perform on its own terms and develop its own sense of creative expression within opera performance.

Robot Opera at University of Sussex

The *Robot Opera* project launched at the Centre for Research in Opera and Music Theatre (CROMT) in 2017.[3] At the time of writing, the project has had a total of two workshops and four public performances, each of which featured live performances with robots followed by discussions (Figure 8.1).

A core team of interdisciplinary researchers (Ficarra, Hopkins, and Kiefer) have led the project from the outset. The first *Robot Opera* event featured two Nao robots with performances of original, composed pieces titled O, *One* (music and libretto by Ficarra) and *The Opposite of Familiarity* (music by Ed Hughes, libretto by Eleanor Knight). *Robot Opera – What's Next?* featured a Pepper robot, a robot cello, a robot cat and two human performers, and *Robo_Po // Robo_Op*, included robot poetry reading and an operatic performance featuring Cleo, a realistic humanoid robot designed and built by Engineered Arts. The *AI & Creativity*

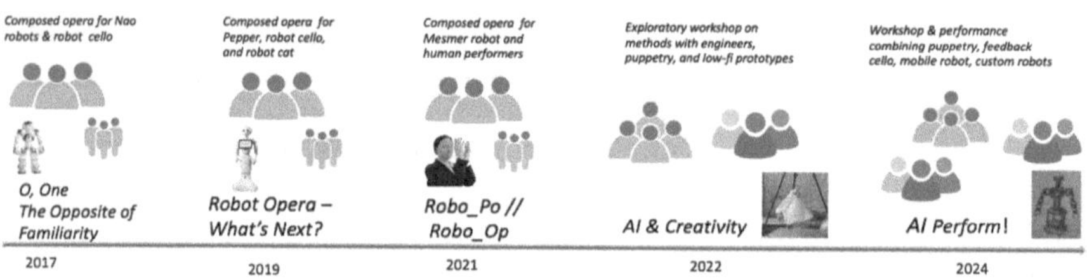

FIGURE 8.1 Timeline showing the development of *Robot Opera* project through to *AI Perform!* The project is led by Evelyn Ficarra and core team of researchers from Sussex University. Figure courtesy of the authors.

workshop was hosted by Elizabeth Jochum at the RELATE Research Laboratory for Art & Technology at Aalborg University, and the *AI Perform!* workshop was conducted at the University of Sussex and included a public performance (2017).[4]

The primary research questions guiding the *Robot Opera* project are: how might a robot sing if it sang *like a robot*, rather than emulating a human voice? How might the robot's own materiality produce an authentic voice? On a more philosophical level, when a robot sings, who exactly is singing? Finally, what values, aesthetic choices, and biases are manifest when a robot sings? Opera as an art form offers unique affordances for our enquiries into embodied AI, as singing and emotional experiences are highly constructed and yet produce authentic emotional experiences.

We know from prior experience with humanlike robots that human operators are at the centre of much decision making. Important factors concerning how a robot looks or sounds, as well as specific design choices such as the options of male/female voices and accents, involve implicit assumptions and decisions that have a tremendous impact on the resulting behaviour and interaction. Sometimes, these design choices are hard-wired into the system and can become constraints when attempting to work with robots in unconventional ways. For example, for the first three performances with humanoid robots, it was clear that none of these robots were designed to sing. The robots' text-to-speech (TTS) software had been optimised for intelligible human speech, but not formatted for singing or other types of vocal expression. We had the option to import pre-recorded vocal material, but we wanted the robot to sing with the same voice it used for speaking.

To do this, we had to find a way to go "under the hood." We first attempted to adapt musical practice to the voice software using a concept called *extended techniques*, where a classical or traditional instrument is explored and played in unusual ways to create new sounds (Tindall-Gibson, 2022). We applied this technique to the robots through software that would allow us to manipulate specific parameters of the speech synthesis, for example, the pitch and rhythm. Ficarra spent hours experimenting with the speed and shape controls for the robot voices, sometimes typing gibberish words that were extremely challenging for the robot to reproduce, with startling and often humorous results. Later, Ficarra extended this approach further to include recorded materials from the robot itself, both its speaking voice and the mechanical sounds created by the robot's motors when moving. Ficarra also visited the Engineered Arts robot factory in Cornwall (UK) to record the sound of machinery as the robots were being manufactured, in an effort to incorporate the robot's materiality into the repertory of vocal sounds and utterances. These sounds were combined with recorded sounds from Loré Lixenberg, a professional opera singer who would perform alongside Cleo, creating a hybrid synthesised voice made of computer, mechanical equipment, and human voice that was tele-operated during the performance. We found that this expanded hybrid vocal world enabled us to vastly extend the robot's expressivity, which greatly influenced our approach to the next phase of the project.

Following the work with commercial robots, we decided to rethink our approach to devise new ways of exploring vocal virtuosity for robot performers. We began by experimenting with training a RAVE model on a corpus of recordings from Lixenberg. RAVE models are variational autoencoders, specifically designed for real-time generative audio. Developed by Antoine Caillon and Philippe Esling, RAVE models are generative audio models trained on large databases of sound (Caillon and Esling, 2021). Once trained, they are able to generate new sounds that are approximately consistent with the sounds on which they were trained, either as a result of feeding live audio into the model or by navigating through the model's internal parameter space. Once trained, it is possible to generate sound either by navigating through the model's latent space or by using sounds to cue output from the model. The model allowed us to experiment with generative singing and vocalisations. We set up an additional mode of control, where the model was cued by output from a Text to Speech engine. We were also keen to explore the new machine learning tools for generating creative content in music, poetry, images, and choreography.

Towards Embodied AI – Workshops on Creativity, Puppetry, and Performance

We organised two workshops on robots, AI, and opera with researchers from Aalborg University and CROMT. The primary goals of the first workshop were knowledge sharing, idea generation, and practical development towards joint collaboration across robots, puppetry, opera, and embodied interaction. We were also keen to pilot test workshop formats and develop tools that would facilitate future collaborations and avoid the parallel work that characterises many art and technology performance projects. Both workshops had a strong focus on voice, and in particular how voices are connected to materiality, body, and movement. We drew heavily on established improvisation techniques across music, theatre, and dance. The workshops were structured into practical working sessions, which involved working in small groups to combine physical materials with software that resulted in low-fi and mid-fi prototypes. Workshop 1 prioritised exploration over technical implementation, and Workshop 2 built on the outcomes of the first workshop, moving from low-fi prototypes to higher-fidelity prototypes in collaboration with musicians and performers. Participants were asked to present concrete exercises, hands-on tutorials, or demonstrations of their artistic or research practice. This could take the form of leading participants in exercises utilising a piece of software, tool, instrument, or group activity, and the focus was on generating a shared vocabulary and "tool box" that could be used in material investigations in the development of an artificial performer. Presentations included Dall-E, poetry generated using ChatGPT, algorithmic music, physical and vocal improvisation, live coding, puppetry, and devising for lyric theatre.

Improvisation

Because our participants came from mixed backgrounds, some of whom had little prior experience with voice, it was important to begin with activities that invited participants to listen, observe, and attend to their own voices and one another's voices and bodies. Drawing on classic improvisation techniques, Hopkins led the group in guided activities such as using bodies as music notation, creating irregular rhythms, lyrical staging, and creative movement that functioned as introductory lessons in musical and spatial composition. Simple exercises demonstrated how everyday sound, gestures, and materials could be developed and transformed from pedestrian activities to more formal, narrative, and performance-oriented actions through attention, composition, and coordination.

Computational Interfaces for Improvisation

The idea behind these tasks was to build a bridge from the analogue improvisation techniques to working with basic computational tools and pre-trained models from computer music through improvised group exercises. Kiefer introduced an interactive voice synthesiser that uses concatenative synthesis, a technique where a collection of sounds is segmented into short chunks and organised according to various musical qualities (pitch, texture, similarity, etc.). Once organised, the player can navigate through these segments, recombining them into new sounds. Other demonstrations were given on the topics of live coding (Tatar et al., 2023), computer music (Bemman and Meredith, 2018), image and text generation using ChatGPT and Dall-E, and a Bidirectional Long Short-Term Memory BiLSTM neural network (Li et al., 2022).

Puppetry, Objects, and Animation

Both workshops included guided activities on puppetry and object theatre. Jochum led puppetry improvisations working with soft, deformable stick puppets made of foam, found objects, newspaper, and other materials. Participants were asked to explore objects and materials individually, then in pairs, and then in groups with their objects and simple puppets, first using movement only (no dialogue) to communicate and explore the sonic features/affordances of the simple objects. Later, participants were introduced to basic principles of animating puppets – *focus*, *weight*, *breath*, and *thought* – in a series of guided activities. As with other forms of improvisation, stories, characters, idiosyncratic behaviours, and scenes began to emerge quickly from these simple building blocks.

Material Investigations

The second part of the workshop involved material investigations and exploratory proto-typing sessions. Participants worked in groups according to which aspects of tools they were most interested in exploring and documented their processes in sketches, photos, and short video clips. Based on our prior experiences developing robot performances, we knew the tremendous value of working with performers early in the process. Professional performers trained in dance and opera were invited to participate and experiment with the low-fi prototypes, which helped initiate important conversations about how performers can work relationally with technology, rather than being forced to adapt to the constraints of the system later on. We have observed that designers and developers can become very attached to their ideas, without considering how the device might be viewed or activated differently in performance by performers not included in the design process.

In the first workshop, three groups produced three unique robotic prototypes based on interaction, music/sound, and material. We named the prototypes *Isolde*, the *Puppet Panorama*, and *Twisty*. All of the prototypes were based on Arduinos and servo motors to activate the movement (Figure 8.2). The *Isolde* prototype was inspired by the foundations of vocal and physical improv exercises – the ability to listen and respond, make a sound based on a sound made by another person (hearing, repeating, vocalising), and move independently while making a sound (gesturing). We adapted Kiefer's Operatic Voice Instrument, using the vocal database based on concatenative synthesis based on previously recorded snippets of vocal improvisation. Concatenative synthesis isn't a new technique, but it has been revitalised by its inclusion in FluCoMA (Fluid Corpus Manipulation Project) framework (Tremblay et al., 2021), with the benefit of new machine listening techniques.[5] The FluCoMA framework was used to analyse around 40 minutes of opera singing and organise the sounds on a 2D plane, so that a performer could trigger the sounds by navigating a path through them using a gamepad joystick. The sound from the voice synthesiser was

FIGURE 8.2 Workshop 1 *AI & Creativity*. Participants formed groups and produced three prototypes of robotic systems for performance based on interaction, music/sound, and material. Clockwise from top left: performer Valeria Rizzo interacts with "Isolde"; Ficarra and Erkut animate the "Puppet Panorama"; Hopkins with "Twisty"; Tatar, Ficarra, and Kiefer interact with the puppet "Isolde." Photo credit: Elizabeth Jochum.

analysed using MFCCs (a feature that characterises the texture of sound), and the resulting coefficients were mapped onto the motion of four servo motors. The "body" was made of lightweight fabric and thin wooden sticks for arms, and we used a tripod to operate it like a marionette.

A musical phrase would trigger the robot to move, causing the movement of the servos to make sound. The motor sounds were picked up by contact microphones attached directly to the servo motors, amplifying the robot's natural, embodied "voice" and linking the robot's physical gesture with vocalisation. The motor sounds were analysed using the Shannon Entropy of low and high frequencies, and these signals were added to the gamepad motion.[6] This formed a feedback loop, creating emergent behaviour patterns in the puppet. None of the movement sequences were very long, which resulted in short, jerky movements (which appeared more fluid because of the fabric). We added a microphone later, and mixed the input signal with the motor sounds. This created two modalities for interacting with the robot, either through navigating the vocal instrument database with the joystick or directly with the voice. This last feature was important because it moved beyond a dyadic interaction and allowed for more performers to interact with the robot in real time.

Building on the ideas uncovered in the first workshop, the second workshop focused on themes of communication, behaviour, and control that linked the robots' physical embodiment with voice (Figure 8.3). Collaborating with the Future Technologies Lab at the University of Sussex, we developed several custom robot performers including a mobile robot with two arms made of servo motors named "Dexter," a robot bird chorus, an augmented feedback cello, and several sonic artefacts equipped with transducers and audio exciters that vibrate rigid surfaces to create sound. Dexter had a wheeled base and traversed the space through teleoperation. It was equipped with an array of speakers and sensors, including a small camera, that enabled the robot to move and respond to the environment through listening and repeating sounds.

Steve Symons and Max Worgan augmented the Dexter robot using AI techniques. The robot carried wireless speakers, controlled from a nearby laptop. Sound was generated with a RAVE model, where the latent space was navigated through a dense multilayer neural network, which was optimised to create mappings from the positions of the servos on the robot. Audio output only occurred when the robot gestured, to emphasise the

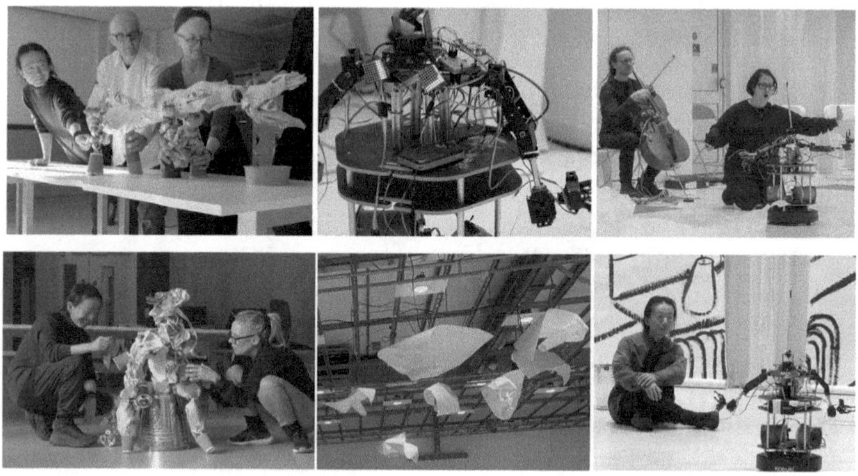

FIGURE 8.3 Workshop 2 *AI Perform!* Explored themes of Embodied AI through robot prototypes that combined puppetry, lyric theatre, and experimental music with topics of communication, behaviour, and control. Clockwise from top left: workshop participants animate a paper puppet; the robot Dexter; Loré Lixenberg and Kiefer on feedback cello perform with Dexter; Shu Yang with Dexter; a chorus of singing birds made of paper and minicell phone vibrator motors; Shu Yang and Cécile Chevalier make a newspaper puppet. Photo credit: Catalina Balan.

robot as an embodied agent that talked or "sang" through movement. The features of the mapping parameters were slowly changed during the performance to give a sense of learning or adaptation to the human opera singer.

The robot bird chorus was a set of eight paper and cardboard sculptures suspended from the ceiling, each equipped with two minicell phone vibrator motors controlled through Arduino that caused the cardboard bodies to vibrate and produced a "chirping" like sound of various, irregular, and random frequencies.

The feedback cello (Eldridge and Kiefer, 2017) employed the concatenative synthesis technique described above, where sound from a corpus of singing recordings was organised on a 2D plane according to similarity and recombined into new sounds. This model was embedded within the feedback loop of the cello, using feature analysis of the instrument sound to navigate through the 2D plane, adding synthesised sound into the circulating signals. Feedback instruments automatically provide a level of unpredictability in performance; even when they are controlled in real time, the human musician is never fully in control of the performance. Feedback adds a life-like quality and a sense of autonomy to the instrument.

The performance was devised collaboratively by Ficarra and Hopkins with input from the entire team. The performance space was conceived as a sonic environment where performers were invited to perceive, attend, listen, and interact with the environment and with one another. The environment contained several sound objects arranged throughout the space, including two traditional visible audio speakers, one audio speaker hidden inside a deconstructed piano, two transducers (one attached to a cello, the other attached inside a large mound of white vellum paper), and the bird chorus. Ficarra deployed manipulated recordings of these objects (cello, birds, paper, piano) played back through the objects using Max MSP, so the objects resonated with sonic traces of their own physicality and previous activity. Using transducers in this way is sometimes referred to as "re-embodied sound" (Goodheart, 2015); the technique also relates to the concept of "sonic ghosting" (Bright, 2020) and resonates with work by Ashley Fure (Fure, 2018). Performers interacted with the environment and sonic artefacts, and the dramaturgy evolved through listening and imitation between human and machinic performers, each vocally and physically mimicking one another to create a partially improvised score. The mimicking practice was a metaphoric reference to the underlying logic of machine learning models.

Discussion

The *Robot Opera* project began as an investigation into the nature of robotic voice and the potential of robots and AI for opera. Frustrated by the constraints of working with the standard, synthesised voices of commercial robots, we focused instead on material and sensorial explorations through generative and embodied performance practice. Admittedly, our prototypes were more rugged in appearance and more difficult to control, but they opened up new directions in our thinking about how we might develop interactive, machine interpreters that are able to improvise and interact with human performers and the environment in real time. Opera involves the convergence of different expressive agents, and often the moving parts are not only visible but also a feature of the performance. These aspects of convergence and artificiality fed into our performance investigations – from the visibly constructed nature of our robots to the collaborative engineering of the performance. We found that approaching the design of robot characters from a relational approach was key to unlocking our collaborative creativity and charting new pathways to work across disciplines. Following the workshops, we conducted structured evaluations, and five themes emerged: participatory design approaches; the importance of the environment; materials matter; challenges and limitations of working with commercial robots and pre-trained models; and the importance of performance to research.

Participatory design approaches. Hands-on activities and having tools ready to work with on the spot were essential for launching group work and creating an atmosphere that put everyone on the same level and encouraged experimentation. The focus was never on technology *per se*, but on knowledge sharing of the diverse expertise in the room. Equal space and value were given to technical and non-technical exercises, and there was sufficient time to explore each of the topics without pressure to immediately programme a robot or work with a pre-selected AI tool. Some of the more technically oriented participants had little experience with opera or theatre, while some of the theatre makers had little programming experience. The hands-on activities established common ground, which enabled participants to explore together rather than dividing up into "tech" and "non-tech" camps. The collaborative nature of the workshop was a pleasant surprise for many of the participants, most of whom had come with different expectations.

The importance of the environment. The physical environment is important for making people feel comfortable working in new ways with new materials. The rooms were flexible spaces that could be rearranged on the fly. It was also important that we were all in the room together, and that the spaces were large enough to provide each group with an ample working area, as well as corners to have some down time, either to work on code, work through an idea, or just relax amidst the flurry of workshop activities. The time scales of working with theatre, choreography, puppet making, and coding vary and can be unpredictable. It was important that participants could still feel connected to the work and process, even while waiting for a set piece to be assembled or code to be debugged.

Materials Matter. All the workshop leaders had prior experience with commercial robots. The decision to work with bottom-up approaches, animating very rough and crude materials (mesh, paper, cardboard, newsprint, foam, tulle, string, cloth, foam, and fishing line) to investigate their movement and sonic/vocal potentials opened up more space to develop new strategies for interaction. Material investigations and puppetry exercises were instrumental in priming participants to think about working with and through material, instead of imposing motion onto an object (a very different approach from programming conventional robots). Access to diverse materials and tools was key to quickly making physical prototypes and realising new ideas on the fly without investing too much time or energy.

Challenges. Bringing disparate research fields together and bridging methods and practice is not without challenges. The most challenging part for us was structuring the sessions to allow for different kinds of creative work, and balancing keeping the experimentation process going while documenting outcomes with a focus on performance. We also found that it takes substantial time to generate a shared understanding and common vocabulary. How sure are we that we have a coming understanding of what a voice is? Is voice something that we make with our bodies, like a cricket does? Or does voice need a mouth? Does voice always imply words and communication, or can it just be sound? Equally important are cultural and disciplinary considerations when it comes to conventional understanding of voice and vocal virtuosity. The "voice" of an instrument is not equivalent to that of a human, and voice in Western opera is characterised quite differently, for example, from voice in Chinese opera, with different aesthetic conventions and vocal techniques for producing sound. Taking time to listen, watch, share, absorb, test, and discuss materials, processes, values, and assumptions can help interdisciplinary teams to generate common references and define creative directions.

Another challenge is how to work with AI tools collaboratively. Although we experimented with Chat GPT4, we found the emphasis on semantic language took us further away from voice and embodiment. In theory, *Dall-E* could be useful for storyboarding or creating mood boards, but it didn't seem terribly relevant to our investigation. When it came to working with AI tools, we primarily used a complex systems approach that supported the interactions between robots, human performers, and the environment. For the Dexter robot, the underlying sound synthesis was a RAVE model, but the focus was not on the generated output but rather on the exchanges between the human performer and the robot. The computer interface Kiefer designed is easy to play with and highly accessible. The content generated by the pre-trained model was far less interesting from a performance perspective than the improvised interactions between the human performer, the pre-trained model, the robot, and the feedback cello that the model enabled, reminding us that the creativity is not native to the tool but in *how we use the tools*.

The Importance of Performance in Performance Research. Both workshops involved formal and informal presentation sessions, which are essential to test out ideas and get feedback during the devising and designing phases. Audience members were from the university community, largely educated and familiar with experimental theatre and music conventions. Inviting the public adds a layer of expectation and emphasis on a finished product, but also helps us to uncover new meanings, and consider the ways that our artistic research connects with broader societal concerns and curiosity surrounding robots and AI.

Conclusion

Robots onstage present new dramaturgical considerations and provide a site for rethinking and reconfiguring the evolving relationship between humans and machines. Opera has proved particularly conducive to experimentation with embodied AI, partly due to its long history with technology and experimentation, and also because of the unique way opera combines artificiality and authenticity. Opera announces its artificiality even as it engages audiences at a profoundly emotional level, with music allowing opera to make this special claim. Music offers forms of creative thinking that help us to imagine AI as an emotional construction as well as a physical one, and to reflect on the artificial construction of emotions in human-robot performances.

Reviewing our work thus far, we have become more knowledgeable about the benefits and trade-offs of working with commercial robots versus more open, configurable systems. Robotics companies invest tremendous resources to design robots with sophisticated capabilities, but these are not always the features needed for creative, responsive performers. For creative teams keen to devise new operatic works, it can be helpful to approach the task of designing machine performers using a bottom-up approach. Working with open, configurable systems can be valuable for refining and deepening a new understanding of the possibilities for musical expression by robot performers, and imagining what robot opera might look like in the future.

Acknowledgements

An early version of this chapter appeared in a special issue of the journal *Concept*. This research was supported by the Sussex Humanities Lab and by Higher Education Innovation Funds administered through the School of Media, Arts and Humanities and the School of Engineering and Informatics at the University of Sussex.

Notes

1 See Sone 2017, Reilly 2019, and Jochum, Millar, and Nuñez 2017.
2 See Burnham 1968 and Kac 1997.
3 Initiated by Evelyn Ficarra, the team of Sussex collaborators has included Ron Chrisley, Tim Hopkins, Ed Hughes, Chris Kiefer, Carol Watts, Kat Sinclair, and others.
4 The *AI & Creativity* workshop held at Aalborg University (2022) included Elizabeth Jochum, Daniel Overholt, Brian Bemman, Chen Li, Dimitrios Chrysostomou, Cumhur Erkut, Evelyn Ficarra, Chris Kiefer, Tim Hopkins, Kivanc Tatar, and Valeria Rizzo. Participants in the *AI Perform!* workshop held at Sussex (2024) included Evelyn Ficarra, Tim Hopkins, Elizabeth Jochum, Steve Symons, Max Worgan, Cecile Chevalier, Loré Lixenberg, Dimitris Kyriakoudis, Chrisantha Fernando, Shu Yang, and Bill Bigge. Andy Philippides, Phil Husbands, and Dexter Sheperd contributed to the robot development prior to the workshop.
5 The Fluid Corpus Manipulation project (FluCoMa) instigates new musical ways of exploiting ever-growing banks of sound and gestures within the digital composition process, by bringing breakthroughs of signal decomposition DSP and machine learning to the toolset of techno-fluent computer composers, creative coders, and digital artists. https://learn.flucoma.org.
6 Shannon entropy is part of communication theory that focuses on the minimum amount of information required for transmitting a signal.

References

Bemman, Brian and David Meredith. "Generating new musical works in the style of Milton Babbitt." *Computer Music Journal*, 42 no.1 (2018): 60–79.

Bright, Danny. "Sonic Ghosting: interrogating place/space/memory through multimodal sonic fracture." PhD diss., University of Sussex, 2020.

Burnham, Jack. (1968) "Systems Esthetics." *Artforum*, 7 no. 1 (September 1968): 30–35.

Caillon, A., & Esling, P. 2021. RAVE: A variational autoencoder for fast and high-quality neural audio synthesis. arXiv preprint arXiv:2111.05011.

Eldridge, Alice and Chris Kiefer. "Self-resonating feedback cello: interfacing gestural and generative processes in improvised performance." *NIME, Copenhagen, Denmark* (2017): 25–29.

Elizabeth Jochum, Philip Millar, David Nuñez (2017). "Sequence and Chance: Design and Control Methods for Entertainment Robots", *Robotics and Autonomous Systems*, 87, 372–380, ISSN 0921-8890, 10.1016/j.robot.2016.08.019

Fure, Ashley. "Ashley Fure | Adam Fure | International Contemporary Ensemble: The Force of Things: An Opera for Objects." 2018. PEAK performances. Accessed 14 May, 2024, https://vimeo.com/223820043

Goodheart, Matthew. 2015. "Matthew Goodheart: Re-embodied Sound and its Implications." CNMAT. Last modified 8 May, 2015. https://cnmat.berkeley.edu/event/2015/05/12/matthew_goodheart_re_embodied_sound_and_its_implications

Ingold, T. (2009). The Textility of Making. *Cambridge Journal of Economics*. 34, 91–102. 10.1093/cje/bep042

Jochum, Elizabeth and Todd Murphey. "Programming Play: Puppets, Robots, and Engineering," In *Routledge Companion to Puppetry and Material Performance*, edited by Dassia Posner, Claudia Orenstein and John Bell, 308–321. Abington: Routledge, 2014.

Jochum, Elizabeth, Philip Millar, and David Nuñez. "Sequence and Chance: Design and Control Methods for Entertainment Robots." *Robotics and Autonomous Systems*, 87 (2017): 372–80. https://doi.org/10.1016/j.robot.2016.08.019

Kac, Eduardo. 1997. "Foundation and Development of Robotic Art." In *Art Journal* (56): 60–67.

Li, Chen, Andreas Kornmaaler Hansen, Dimitrios Chrysostomou, Simon Bøgh, and Ole Madsen. "Bringing a Natural Language-enabled Virtual Assistant to Industrial Mobile Robots for Learning, Training and Assistance of Manufacturing Tasks." *IEEE/SICE International Symposium on System Integration, Narvik, Norway* (2022): 238–243.

Machover, Todd. "Death and the Powers. 24 September 2010." World Premiere: Salle Garnier, Monte Carlo, Monaco. Directed by Diane Paulus, Conductor: Gil Rose. American Repertory Theater, 2010. https://opera.media.mit.edu/projects/deathandthepowers/media.php

Marynowski, Wayde. *Robot Opera*. YouTube. 9 December 2015. Video. https://www.youtube.com/watch?v=eOgKzrsgb2s&t=87s

Reilly, Kara. "Robots and Anthropomorphism in Science Fiction Theatre: from Rebellion to Domesticity and Back Again," In *Media Archaeology and Intermedial Performance Deep Time of the Theatre*, edited by Nele Wyants, 193–210. London: Palgrave Macmillan, 2019.

Savery, Richard, Lisa Zahray, and Gil Weinberg. "Shimon Sings-Robotic Musicianship Finds Its Voice." In *Handbook of Artificial Intelligence for Music*, edited by Eduardo Reck Miranda, 823–847. Cham, Switzerland: Springer, 2021.

Schaeffer, Simon. "Enlightened Automata." In *The Sciences in Enlightened Europe*, edited by William Clark, Jan Golinski, and Simon Schaffer, 126–165. Chicago: Chicago University Press, 1999.

Sone, Yuji. *Japanese Robot Culture: Performance, Imagination, and Modernity*. New York: Palgrave Macmillan, 2017.

Tatar, Kivanc, Kelsey Cotton, and Daniel Bisig. "Sound Design Strategies for Latent Audio Space Explorations Using Deep Learning Architectures," in *Proceedings of Sound and Music Computing, Stockholm, Sweden* (2023).

Tindall-Gibson, Annie Rose. "Extended Techniques in the Repertoire of the Intermediate Learner." *The American Music Teacher*, 72, 1(2022): 21–23.

Torpey, Peter (2012). "Digital Systems for Live Multimodal Performance in *Death and the Powers*." *International Journal of Performance Arts and Digital Media*, 8(1): 109–123.

Tremblay, Pierre Alexandre, Gerard Roma, and Owen Green. "Enabling Programmatic Data Mining as Musicking: The Fluid Corpus Manipulation Toolkit." In *Computer Music Journal*, 45(2) (2021): 9–23.

9

WHAT DOES IT MEAN? ON PLATFORM-INVARIANT "BODY-LANGUAGE" DICTIONARIES

Amy LaViers

1 Introduction

This chapter discusses the possibility, opportunity, and challenges of cataloging a broader swath of human movement – postural and gestural shifts that do not produce vocalisations recognised as language or music – or so-called body language. The result of this task would be a rigorous cataloging of so-called body language – the postural and gestural shifts involved in human behaviour that are known to be significant in signaling between humans (Aviezer et al. 2012). Such a catalog would open up nonverbal interaction with machines, including software assistants, physical robots, and other situated automation.

If we could arrange the wide array of the bodily patterns of the eight trillion people on the planet onto a spectrum or grid, we would notice patterns that result from (and likewise create) shared culture (Dils and Albright 2001). This organisation would not be static, it would evolve over time as both the number and composition of humans (and their activities) on the planet change.

Regional differences in movement patterns would appear. There would be an area of head tilts in the coronal plane, about the sagittal axis of the head specific to gestures of agreement used in parts of Southeast Asia. There would be an area of intricate but strong maneuvers of the fingers of fishermen slipping wet, wriggly bait on a hook. These particular bait-on-hook patterns would be concentrated in rural areas where bodies of water with legal fishing are more common, and subregions within this activity would be regional: specific to the areas where particular bait types are

DOI: 10.4324/9781003475972-11

favored by the local fish – worms versus leeches versus crickets, say – all whose bodies require slightly different movements to ensnare.

There would be a region that highlights where and which bodies practice ballet – versus the many areas where the form is not commonly taught. Within that region, Russia in the 1800s, with its stern Vaganova method, will be distinct from the Balanchine style that has grown dominant in parts of New York City today. Cell phone users would form a region, grouped by operating system, with iOS and Android users forming the largest lobes – their subtly distinct proprietary swiping, pinching, and flicking gestures sorting into two adjacent but distinct regions. These patterns cut across functional needs and expressive styles, body sizes, and, maybe most importantly, cultures: when, where, and among whom the movements occur greatly shape their expressed form.

Another region in our multidimensional movement spectrum that is especially easy to track and coordinate with culture is the region containing the movements that produce spoken language. These would be grouped according to which language the speaker is speaking: a feature closely associated with culture – both resulting from it and creating it at the same time. New clusters on the grid emerge and recede along with the development and extinction of spoken languages, dialects, and creoles.

For example, the United States in the second millennium would show up as a region of English speakers engaging in the jolting, Germanic forms of this language, accented by words from languages from around the globe. States like Texas and Florida where a growing body of citizens also (or instead) speak Spanish would take on the softer, more fluid forms and quicker pace of that language's flow and speed. Europe and China would be dotted with its many distinctive tongues, packed into tight, region-specific clusters – with an ever-present layer of English and Mandarin, national languages often spoken alongside older, native dialects – relics of a time when communication and travel were more limited, and languages tied more tightly to their regions of origin and use.

And, in every case, we can imagine a situation where someone observes the movement, scratches their head (if the observation is happening in the spaces and times that utilize this idiom to describe an action of puzzlement), and asks: what does it mean? The answer depends on the context of the action.

The organisation we have imagined here is vast and as yet beyond our systematic understanding to create. Some shadow of it is beginning to be captured in data, for example, our travel patterns in vehicles as detected by built-in GPS units, the locations of our fingertips on touch screens, the acceleration of our wrists wrapped in fitness trackers, and even our posture in certain contexts as tracked by extensive software suites and depth cameras like the sensor systems on autonomous and semi-autonomous vehicles, but efforts to make sense of it in a systematic way – especially for nonverbal behaviour – have come up short.

Lacking a holistic, data-based system for organizing this phenomenon, one way we might catalog and organize behaviour is as a dictionary. Many dictionaries – and their ever-changing styles and editions – have been written since the late 19th century to document the verbal movement behaviours, and corresponding intents, comprised by English-speaking people (Ogilvie 2023). Thus, this chapter considers the process of creating such a document dissecting nonverbal behaviour. The challenges highlight gaps in the understanding of bodily movement. Imagining such a document paints opportunities for creating new relationships between human and robot bodies. And, the limits of current-day dictionaries of vocalisation, that is, their inability to capture the different *ways* a speaker might utter a particular word and connect them with the intended meaning, suggest similar limits on what this interaction might look like for nonverbal behaviour.

In Section 2, I outline technology (notation) for describing movement associated with vocalisation and audio production, highlighting the relationship between letters that form words and bodily action and comparing tablature and modern music notation. Section 3 outlines systems for describing the movement of human bodies more broadly, focusing on the pattern of gestural and postural shifts in bodies; this section includes a mature system for documenting movement as performed by trained and able-bodied dancers as well as emerging systems that need not necessarily be related to the human form. Section 4 describes a potential design of "body-language" dictionaries, codifying postural and gestural bodily shifts, with three key features imitated from traditional dictionaries. Section 5 highlights systems for robot programming (including some that explicitly leverage movement notation), which also serve as a record of the movement of a particular body, and human-robot or human-AI interaction; importantly, the structure of these systems and any potential "body-language" dictionaries, creates a cycle of influence that will shape the aims of bodies both natural and artificial for centuries to come. The chapter concludes with impressionistic ideas and conclusions about this space for trying to understand human movement and to design novel technological systems to interact with it harmoniously.

2 Notating Vocalisation (in Spoken Language and Music)

Human language is typically produced through movements of our vocal cords, diaphragms, mouths to create vocalisation (Ghazanfar 2008), and often hands or some technological system, for example, a person who uses a speech-generating device (Rispoli 2010). In every case, it is produced through bodily movement, and there is an extensive and widely used notation system to guide the reproduction of these movements across many bodies: written language.

It is not known how long humans used spoken language – estimates range widely, between 40,000 and 2,000,000 years ago (Janson 2012, 5–6) – but written language is a notably more recent development, beginning around 5,000 years ago (Janson 2012, 54). Across this time, many distinct languages and their corresponding cultures have emerged and extinguished; that is, human cultures find and discard – and are *defined by* – movements for communication (Janson 2012). These movement patterns (and their notation) are central fixtures of ever-evolving human cultures.

Around 1,000 years ago, modern music notation was invented in medieval Europe, replacing tablature – a feat I have written about as shifting from platform-specific to platform-invariant notation (LaViers and Maguire 2023, 57). This system for identifying pitch (Guido of Arezzo, 1030) and duration (Leoninus of Notre Dame, 1180) came into convention in Western culture for capturing the patterns in religious hymns of the time (Kelly 2014). This notation had an important difference from tablature: the same score gave instructions for human vocal cords, pianos, lutes, clarinets, cellos, and a wide array of instruments known to the Western world (and would later describe behaviour on yet to be encountered or invented instruments that produce tonal sounds). The goal of these medieval monks was simple: they wanted to be able to "learn songs which they had never heard" (Kelly 2014, 65). But they developed something broader: platform-invariant (Jang Sher et al. 2019) instructions for creating rhythmic patterns with tonal sounds that would last for centuries (and counting).

As useful as written language and notated musical scores have become, it is important to take note of what they *fail* to represent. Any abstract system of notation must account for many distinct performances of a given movement phenomenon. That is, a notational system must unambiguously identify many distinct "compliant performances" (Goodman 1976, 178). In the case of music, the tonal notes invented by Guido threw out much of musical performance, for example, timbre and melodic shape (Kelly 2014, 79–80), creating many ways to successfully portray the same score.

Likewise, dictionaries record words along with their pronunciations, but there is a wide range of performances, or physical utterances, of each word. A speaker may vary their phrasing, changing which part of the word they emphasize, for example, through increased (or decreased) vocal volume (LaViers and Maguire 2023, 152–158). Or, a speaker may create inflections through changes in their tone or breathing patterns (Gates 2022, 39). A speaker may successfully pronounce the word in a range of different, regional accents. Thus, there is much about the performance of a word that neither musical scores nor written language capture.

Even the outline of this development of music notation follows my own cultural situation in the West, using words and instruments common to my own English-speaking experience listening to Western music styles. Like language, dictionaries and music notation are cultural artifacts. These particular artifacts are robust sources of instructions on how to move our bodies in a variety of contexts – that is, they are movement *scores*, as discussed in the next two sections.

2.1 Western Sheet Music as Movement Scores

The basic unit of modern sheet music is the note. It specifies the pitch and duration of a sound (which can be generated by the vocal cords or any one of many instruments that produce tonal sounds). There are 12 different tones (C, C sharp/D flat, D, etc.), which repeat across the 10 or so octaves in audible sound range, and at least 9 different durations (whole, half, quarter, etc.). Notes are arranged on two staffs (the bass and treble clefs) across eight octaves. Pitch is indicated by the vertical location of these two staffs, and duration is indicated by the shading and styling of the note. Notes are arranged in the order they should be played from left to right across the page. Staffs also contain additional information that guides the playing of notes: the meter of the music is typically given at the start of the score; quality of note playing may be indicated by notes like *staccato* (sharp, detached notes) and *legato* (smooth, connected notes); rests indicate the absence of notes; and crescendos and decrescendos indicate growing and diminishing volume – to name a few.

How do these familiar notes translate to movement instructions? This depends on the instrument one is playing. For example, to play a quarter C note: a pianist finds a C note key (marked by being to the left of the paired black keys and two keys to the right of the trio of black keys) and depresses it with any of their fingers for four counts (or beats of a metronome, say) at a prescribed tempo. The pianist may also use their foot on the right foot pedal to release their finger but let the sound of the note continue, cutting off the typical damping action of the piano's internal mechanism. To sing such a note, the singer needs only to control their breath and the constriction of their vocal cords. Thus, many distinct movements, requiring multiple body parts, on many distinct instruments satisfy the score.

2.2 English Dictionaries as Movement Scores

Dictionaries are typically used as reference material in writing and reading (where no audible noises need to be created to understand the meaning of the symbols in a dictionary). But, each entry, nonetheless, provides unambiguous instructions on how to move, and, as with music notation, many different physical phenomena satisfy the score.

The basic unit, then, of the dictionary score is a letter. Letters are grouped together in complex ways that change the pronunciation (or corresponding movement) for each unit. Letters in English texts are read left to right, such that the sound corresponding to the leftmost letters of the word comes before those corresponding to letters moving right. Entries are then ordered in the book this way: all of the words with the same first (and then second, etc.) letter are grouped together. An arbitrary ordering of the letters (A-Z) determines the ultimate sequence.

The English letters (or pronunciation guides) indicate the order of sounds – and corresponding movements that need to be made in sequence from left to right. Thus, from a movement perspective, the words grouped together start with a similar movement. Words like "antigravity," "antihelium," and "antihero," which all start with the letter grouping (prefix) "anti-" start with the same movements of the body and come in order together in the dictionary. However, words with common ending letters (suffixes) are spread around throughout based on their opening letters; so, words with the same ending movements are not grouped together in the dictionary.

Most entries are associated with multiple meanings. Often the way the English word is said is better aligned to one meaning over another. For example, the word "punch" has at least two common meanings as a verb – "strike with the fist" or "press (a button)" – and as a noun – "the strength or power needed to do something" or "a fruity beverage, often mixed with alcohol for a party." The manner that the word is stated may give clues as to which meaning is intended (as do other contextual clues). An excited, relatively loud statement of "punch" versus a softer, more even delivery may distinguish each case: the violent action and the notion of gumption better align to a more energetic delivery than pressing a button or a beverage.

3 Notating Gross Bodily Movement (Posture and Gesture)

In Section 3, I shift gears from vocalisation to consider gestural and postural patterns. Humans have been representing such shifts of our body in the environment since the days of cave drawings. Remnants of this representation are found in cave drawings dating back to over 20,000 years ago (Morriss-Kay 2010). Dark stick figures clustered in particular spatial arrangements and bodily postures give a sense of that early life but contain an unknowable interpretation. Central lines of a figure indicate something about spinal action and the degree of flexion or extension in the various vertebrae. Figures in a line versus a circle indicate a different physical arrangement in space (of multiple bodies across space) or in time (if depicting the same body over time). Despite this long history, the development of codified systems for representing situated postural and gestural behaviour of humans remains an open and evolving task.

There are, however, several codified symbolic notation systems that attempt to represent broader postural and gestural shifts of the body. Each builds on concepts and theories from choreography and somatics, such as the BESST System of movement analysis (LaViers and Maguire 2023), which will be the basis of the dictionary design proposed in this chapter. The BESST acronym is developed from five distinct components of movement, which can be briefly explained through what questions the elements in each component attempt to answer about movement: Body (answering "What is moving?"), Effort (answering "How is the movement performed?"), Shape (answering "For whom is the movement performed?"), Space (answering "Where is the movement performed?"), and Time (answering "When is the movement performed?"). These five categories contain elements represented by symbols and are used to drive several styles of notation, including vertical motif (Guest 1992), Labanotation (Guest 2014), and motif on a novel staff for translation to artificial bodies (LaViers and Maguire 2023).

3.1 Platform-Specific Notation (Concrete, on Able Human Bodies)

Systems of movement notation are often specific to a particular style or canon of movement: Fuelliet-Beauchamp notation was developed to record French court dances and Benesh notation for Western classical ballet. Moreover, specific choreographers often develop their own system or shorthand for their own work. One system that has found use outside of the dance style and beyond the particular choreographer who developed it is Labanotation (Guest 2014), which has been used in animal movement studies (Fagan 1997), dance archiving (Laemmli 2016; Frazier 2021), physical therapy (Gómez 2015), human-computer interaction (Altakrouri 2013), and robotics (Abe 2017; Salaris 2017). Labanotation uses a staff much like music notation: lines with spaces in between that can be divided into measures (or phrases) by bisecting lines. However, symbols move up the page, from bottom to top; often when notating dance, the system's primary usage as codified by Guest (2014), the musical score will be appended, rotated 90 degrees counter-clockwise from its usual presentation. Rather than corresponding to a generalised abstraction like pitch, these lines create dedicated space for body parts, for example, left leg or head, to be annotated.

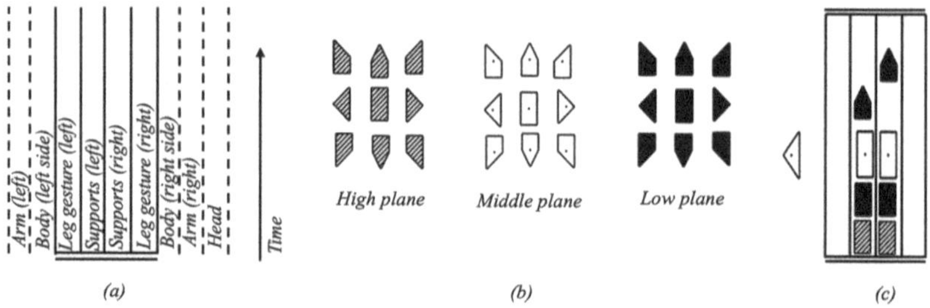

FIGURE 9.1 Labanotation and Space Component symbols: (a) Labanotation staff; (b) Symbols from the Space Component; (c) An example score describing: a relevé (both rising on both feet), then a plié (bending both knees with feet flat), coming to flat feet with both arms out to the side (an action which takes more time than the others in the phrase, as indicated by these middle-level symbols longer vertical extent), taking a step with the left foot, taking a step with the right foot, and then moving the head up, forward, and to the right (all at once). (Credit: Amy LaViers).

The elements of the BESST System critically underpin forms of movement notation. The original symbols used in Labanotation were from the Space Component of the BESST System. These give spatial pulls for each body part, as shown in Figure 9.1. Additional symbols from other components – for example, basic body actions, relationship bows, and shape and effort qualities, which evolved later – can be utilised in Labanotation scores as well. These symbols add further detail and information about the movement expressed. In general, it takes years of training to use this system with competency, and even after that, many variations in annotation style will remain.

3.2 Platform-Invariant Notation (Abstract, on Many Bodies Both Natural and Artificial)

Forms of notation that have been developed after Labanotation may be more appropriate for contemporary use in relation to technologies like robotics and AI. In a more modern usage of the concepts that underpin Labanotation, a notation scheme called motif, symbols are not used on the detailed, platform-specific staff of Labanotation, but instead, a single column, with option columns for modification, describes the main action of a movement idea. Here symbols from the Body, Effort, Shape, Space, and Time Components can be used interchangeably as the focus of the score – and combined and layered to describe more detailed ideas (Guest 1992). This system, thus, tends to favor exchange and expression across bodies, especially in robotics (LaViers and Maguire 2023, 243–244).

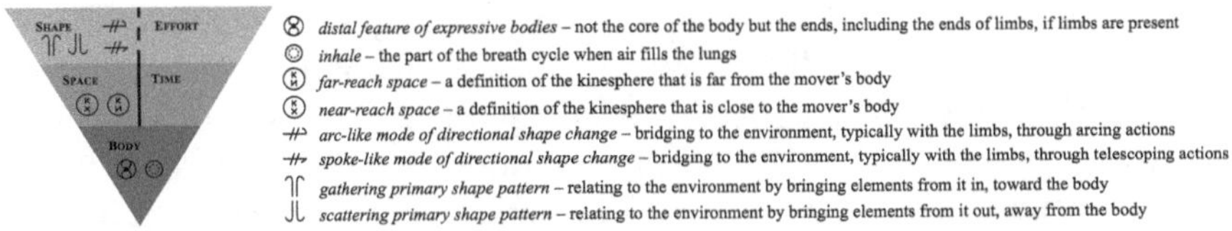

FIGURE 9.2 A subset of Body, Space, and Shape Component symbols used in this chapter from a larger open-source symbol set (LaViers and Maguire, 2023). Definitions are provided at the right, and an arrangement of the BESST System components (and corresponding symbols of this set) is shown at the left. (Credit: Amy LaViers).

By letting go of many more details, this abstraction is closer to modern music notation and written language, then, and will be our target for codifying "movement dictionaries" in the next section. This is because we want each entry to conform to the tradition of dictionaries – and thus consist of notation that can describe the same action on many moving platforms, for example, a human with 1 leg, a robot with 14 limbs, and a virtual assistant shaped like a paperclip. A small subset of these symbols (those used again in the next section) and a representation of how they relate is provided in Figure 9.2.

4 Designing "Body-Language" Dictionaries

A dictionary that catalogs potential meanings of changes in bodily posture and gestures must denote bodily action through an abstract symbolic system – just as traditional dictionaries do for the subset of bodily movement that is vocalisation. Then, we need a method for ordering the entries. Rather than establishing an entirely arbitrary (or alphabetical) convention; this ordering will be based on structural ideas about movement. Finally, we must continue the tradition of enumerating multiple possible meanings. This section discusses each of these aspects of such a catalog in turn.

4.1 Designing the Entries

Dictionary entries are comprised of letters and corresponding phonemes – defined by Oxford Languages as "the perceptually distinct units of sound in a specified language that distinguish one word from another" (OED 2024). These symbols identify the desired phonemes – almost uniquely – and, for any literate speaker fluent in a particular language, unambiguously.

On the other hand, plenty of ambiguity exists around any particular entry's *meaning*. For example, in the case of homophones, the same sound means (and is often spelled) differently. In decoding the meaning of a particular utterance, this challenge must be met using context clues and considering all of the many potential meanings for a given word. However, when viewing words as movement instructions, this does not degrade the clarity of the system. Simply: two distinctly spelled words indicate the same bodily movement.

Each specific utterance of a word is a slightly different movement phenomenon. Differences in, for example, the time it takes to say a word, the tone, and the emphasis will all comprise a successful reading of a given word. Ignoring the special case of homophones, this creates a one-to-many relationship between words and utterances: many utterances can be matched to the same word.

We need a way to represent movement instructions with similar clarity. It is okay that there will be some latitude in expressions for each entry. That is, not every movement corresponding to a given entry will be exactly the same. Latitude may be offered in which body part completes the expression, what spatial plane, what quality of movement, and this fits our experience of movement as well: slight differences in, say, a "thumbs up" gesture in a Western context do not significantly change our perception of the action (nor the meaning of the symbolic token).

I do not know the full set of "phonemes" – or *perceptually defined* "movemes" (Bregler 1997) – of bodily movement. However, I do know some very clear candidates from the BESST System.

The Effort component has long been sought for movement analysis and generation (LaViers and Egerstedt 2012, Knight and Simmons, 2014, LaViers and Egerstedt 2014). However, this aspect of movement is arguably not present in the annotations of a dictionary. Consider the many performances of a given word, for example, "hello." We can use punctuation and alternate spellings to denote differences in performance of the word, which do not impact its entry in the dictionary, for example, "hello!" "hello?" and "hellooooo." These differences in inflection, emphasis, and performance may be explained by effort qualities and phrasing (part of the Effort and Time Components, respectively). Indeed, these differences contribute to the intended meaning, but they are not reflected in dictionary entry spellings. Thus, the entries in this cataglog of posture and gesture will not use symbols for effort qualities or phrasing.

Another aspect which is not declared in a traditional dictionary entry is the element of "body part." Indeed, each entry can be performed by the vocal cords (as in spoken English) or the hands and upper body (as in American Sign Language). Instead, each entry can give movement instructions to disparate body parts on distinct platforms. In this sense, each entry is platform-invariant. I aim to retain this feature in the "body-language" dictionary presented here as well. Thus, entries should avoid naming a specific body part to recognize the varieties of ways and bodies that can engage in nonverbal communication.

This leaves concepts in Body, Space, and Shape Components to notate each entry of the dictionary. (A full list is provided in the next section.)

4.2 Ordering the Entries

In a traditional dictionary, temporal sequence defines the entry order. Time progresses left to right for each letter of the English language in the entry, and words with common prefixes are grouped together, leaving common suffixes in separate sections of the book. In movement notation, however, it is often convention for time to advance bottom to top, with movements that come before (prefixes in our entries) placed below movements that come after (suffixes in our entries).

Moreover, each "letter" is more complex than letters used in words – because it needs to be built from multiple symbols spanning the Body, Space, and Shape categories. As shown in Figure 9.2, the concepts used to notate entries of this dictionary build on each other, with body as the basis (at the bottom of the inverted pyramid in Figure 9.2) and Space and Shape higher-order ideas (moving up the inverted pyramid in Figure 9.2).

Thus, the dictionary will use a two-layered hierarchy to determine the order of entries. First is temporal order (as with words in a traditional dictionary), which is tracked vertically, with the first action in a sequence taking the bottom position of a stack. Second is ordering these movement elements within that first unit of the entry by following the BESST System as outlined in a recent textbook on movement analysis (LaViers and Maguire 2023, 297–310) where concepts in the Body Component underlie the Space and Time Components, which underlie the Shape and (omitted in this catalog) Effort Component.

The list and order of elements used in the design presented here are:

- Body Component
  - Breath and sensing
  - Basic body actions
  - Features of expressive bodies
  - Body parts (platform-invariant)
    - Center of Gravity
    - Center of Levity
  - Patterns of body organisation
- Space Component
  - Zones and planes

    - Spatial pulls
    - Reach space and pathways
- Shape Component
    - Still shape forms
    - Primary shape patterns
    - Modes of shape change
    - Shape qualities

4.3 Enumerating the Possible Meanings of Each Entry

As with different ways to say the word "punch" (using the movement of one's body) described earlier in this chapter, each movement may be performed with different style. By changing elements not notated in this entry design, for example, effort quality and phrasing, a mover may indicate additional nuance. The bandwidth of this nuance should be communicated through multiple possible meanings enumerated.

These enumerated meanings will be limited to a particular time, place, and group of people (a culture). Each entry may have a completely different set of associated meanings (or be entirely nonsensical) for a different culture. What remains to be seen is how narrowly such a culture should be defined. Could it be that a single dictionary could be useful across the West? Or the United States? Or petite able-bodied female adults with training in ballet and yoga who use Apple laptop computers (with a touchpad, never a mouse) and mobile phones (with a 6.1-inch touchscreen) in Philadelphia in the early 21st century, etc.?

4.4 Presenting a Possible Page

The resulting pages might look like the example in Figure 9.3. Each entry may vary in length in two dimensions: vertically (entries with one and two movements in sequences are shown in Figure 9.3) and horizontally (movements with two and three symbols to describe it are shown in Figure 9.3). The order of the entries is created in a systematic order, with movements clustered according to common Body Component elements. Several instructive (not exhaustive) possible interpretations (or meanings) of each action, often associated with additional modifying symbols, are presented for each entry as well. When used in context, elements from the situation, environment, or personality of the mover will help clarify which usage is intended.

Now that I have discussed the basic premises for a "body-language" dictionary, in the next section, I will outline the relationship of such a dictionary to novel and emerging technological systems.

An arc-like mode of directional shape change in far-reach space by the body's distal edge(s) | 1. gesturing for another body to come nearer (when modified by primary shape pattern of gather); 2. gesturing for another body to go further away (when modified by primary shape pattern of scatter); 3. finger-painting on an upright canvas with strokes; 4. brushing dust or other debris off a surface; 5. | typically involves a spatial pull directing the environmental location of the object of the action

A spoke-like mode of directional shape change in far-reach space by the body's distal edge(s) | 1. pressing buttons; 3. finger-painting on an upright canvas with dabs; 4. feeling in the dark for an object (or hand/foothold) of interest; 5. punching or chopping with limb | may be modified by an effort quality to distinguish between some of these ideas

A spoke-like mode of directional shape change in far-reach space, drawing into near-reach space by the body's distal edge(s) | 1. yanking the rope of pulley; 2. gesturing for another body to come nearer with emphasis; 3. a common, decorative action in a cheerleading routine | may be modified by an effort or shape quality

An arc-like mode of directional shape change in near-reach space by the body's distal edge(s) | 1. brushing something off the mover's own body; 2. massaging sore muscles; 3. twiddling one's thumbs | may be modified by a relationship bow to indicate contact with environment or self

An inhale accompanied by bodily action in far-reach space | 1. stretching; 2. celebrating; 3. reaching with intensity | may be modified by a concept in Shape Component to indicate relationship with environment

FIGURE 9.3 A dictionary page mockup showing five possible entries, each with multiple meaning enumerated and ordered according to the system outlined in Figure 9.2. (Credit: Amy LaViers).

5 Machines Powered by (and Powering) Movement Dictionaries

The ubiquity of written language obscures some of the limits of such tools as AI-powered internet searches. For example, when using a tool like ChatGPT-4, queries are provided and answered through the use of symbolic strings (words, sentences, and paragraphs) that are meaningful to human users. The tool cannot exist without this underlying notation. Thus, dictionaries are cultural artifacts that already power AI including text-to-speech systems, the internet, large language models, and generative AI assistants. As dictionaries change, so too will the behaviour of these tools.

For so-called body-language, the path will be the same. Textual description cannot supplant the need for segmentation and classification of the ongoing postural and gestural changes of human movement. What are the tokens in this nonverbal exchange of information? Just as in vocalisation (where distinct languages are used within different cultures and each token has many possible meanings), this answer depends on culture and context.

The codification of nonverbal movement will also impact the physical movement machines can and will exhibit in human-facing environments. In robotics movement is generated directly from programs that specify concrete (not abstract), defined instructions to a specific machine. However, physical inconsistences between units or degradation of parts over time also create a sense of variability and liveness to each machine action – only some of which are perceptible to human beings. Thus, a sense of scoring or "performance" is inherent to the operation of these devices when viewed through the lens of human perception and experience.

The intersection of movement notation and machine programming is then a rich field full of corporeal artifacts of elements of the shared culture of today. Affordances designed by human engineers and offered by our machines create movement of both human users and their machines. For example, consider the following cultural movement practices encoded in technology:

- "Reply All" encodes an alien movement of transistors on servers, producing a particular distribution of information (sending an email to everyone copied on a message rather than just the sender).
- "Thumbs up" represents a seemingly familiar movement of the distal edges of limbs and connects users to content on social media (through "Like" buttons).
- "Swipe" as a gesture of the distal ends on a touch screen; in some contexts, this action means to delete an email while in others it offers the opportunity for connection with another user (as on dating apps).
- Choreographic options encoded for drone movement designers (Schoellig et al. 2014).

Humans will accommodate machines as they are built, taking on the necessary behaviours to utilize the tool. But the examples above show how behaviours are salient for their native time, place, and people – and not beyond. Actions deemed to be universal and anointed as interaction tokens in machines can ossify a culture and alienate it from others. For example, swipe, pinch, zoom, and "like" (thumbs up) have all taken on new meanings and uses in my culture in my own lifetime – due in large part to the development of tools that utilize these actions. That is, their entries in a "body-language" dictionary would have needed to be specific to my location on Earth and updated due to the advent of the iPhone and Facebook.

I want to end with a vision of artificial embodiment – of systems that can interact in a native way with nonverbal tokens. These tokens are not meaning-imbued universal symbols, but cultural and context-specific tools that are used for varied communication (like words, with many possible meanings enumerated in a dictionary). This idea is counter to much research on body language: although the importance of posture and gesture has been robustly measured (Aviezer et al. 2012) and some sounds have been found to have cross-species correlations (Filippi 2017), researchers looking to "comprehend the meaning of body postures" (Raindel et al. 2021) are like dictionary designers looking for a singular meaning for entries (tokens like words). This is an unsuitable approach because that is not how language is used – at least not today. Rather, as there are thousands of dictionaries for vocalisation, collections for each spoken language that have changed over time with multiple definitions for entries, we need many codifications for the broader movement of the body. I predict we will need *millions* of "body-language" dictionaries. As interactions with machines grow beyond our distal edges (our fingertips) to include estimates of postures, gestures, and complex, coordinated activities, it is all the more urgent that these systems support the full, varied, shifting, and layered complexity of human movement.

Acknowledgments

This chapter is partially supported by an AI infrastructure grant from the National Science Foundation (NSF award #2234195), and the chapter has been bolstered through conversations related to developing a notation and annotation system for that project with fellow Certified Movement Analysts (CMAs) and/or dancers who work in science and technology: Dr. Alexandra Bacula, Cat Maguire (who offered helpful comments and corrections for this article), Dr. Si-Hyun Yoo (who offered helpful comments and corrections for this article), Dr. Tal Shafir, and Rachelle Tsachor.

References

Abe, Naoko, Jean-Paul Laumond, Paolo Salaris, and Florent Levillain. "On the use of dance notation systems to generate movements in humanoid robots: The utility of Laban notation in robotics." *Social Science Information* 56, no. 2 (2017): 328–344.

Altakrouri, Bashar, Jan Gröschner, and Andreas Schrader. "Documenting natural interactions." In *CHI'13 Extended Abstracts on Human Factors in Computing Systems, Paris, France*, pp. 1173–1178. 2013.

Aviezer, Hillel, Yaacov Trope, and Alexander Todorov. "Body cues, not facial expressions, discriminate between intense positive and negative emotions." *Science* 338, no. 6111 (2012): 1225–1229.

Bregler, Christoph. "Learning and recognizing human dynamics in video sequences." In *Proceedings of IEEE Computer Society Conference on Computer Vision and Pattern Recognition*, pp. 568–574. IEEE, 1997.

Dils, Ann, and Ann Cooper Albright, eds. *Moving History/Dancing Cultures: A Dance History Reader*. Wesleyan University Press, 2001.

Fagan, Robert, Jan Conitz, and Elizabeth Kunibe. "Observing behavioral qualities." *International Journal of Comparative Psychology* 10, no. 4 (1997).

Filippi, Piera, Jenna V. Congdon, John Hoang, Daniel L. Bowling, Stephan A. Reber, Andrius Pašukonis, Marisa Hoeschele et al. "Humans recognize emotional arousal in vocalizations across all classes of terrestrial vertebrates: evidence for acoustic universals." *Proceedings of the Royal Society B: Biological Sciences* 284, no. 1859 (2017): 20170990.

Frazier, Mara Pegeen. "Labanotation is creative: how a systems perspective reveals generativity in dance notation and its archives." *Journal of Movement Arts Literacy* 7, no. 1 (2021): 105–131.

Gates, Linda. *Voice for Performance: Training the Actor's Voice*. Rowman & Littlefield, 2022.

Ghazanfar, Asif A., and Drew Rendall. "Evolution of human vocal production." *Current Biology* 18, no. 11 (2008): R457–R460.

Gómez, Miguel A. "Rudolf von Laban's Labanotation: the origin of notational analysis methods in sport sciences." *RICYDE. Revista Internacional de Ciencias del Deporte* 11, no. 39 (2015): 96–99.

Goodman, Nelson. *Languages of Art: An Approach to a Theory of Symbols*. Hackett Publishing, 1976.

Guest, Ann Hutchinson. *Motif at a Glance*. Language of Dance Center, Inc. 1992.

Guest, Ann Hutchinson. *Labanotation: The System of Analyzing and Recording Movement*. Taylor & Francis. 2014.

Janson, Tore. *The History of Languages: An Introduction*. Oxford University Press, 2012.

Kelly, Thomas Forrest. *Capturing Music: The Story of Notation*. W. W. Norton & Company. 2014.

Knight, Heather, and Reid Simmons. "Expressive motion with x, y and theta: Laban effort features for mobile robots." In *The 23rd IEEE International Symposium on Robot and Human Interactive Communication*, pp. 267–273. IEEE, 2014.

Laemmli, Whitney Elaine. "The Choreography of Everday Life: Rudolf Laban and the Making of Modern Movement." PhD diss., University of Pennsylvania, 2016.

LaViers, Amy, and Magnus Egerstedt. "Style based robotic motion." In *2012 American Control Conference (ACC)*, pp. 4327–4332. IEEE, 2012.

LaViers, Amy, and Magnus Egerstedt. "Style-based abstractions for human motion classification." In *2014 ACM/IEEE International Conference on Cyber-Physical Systems (ICCPS)*, pp. 84–91. IEEE, 2014.

LaViers, Amy, and Catherine Maguire. *Making Meaning with Machines: Somatic Strategies, Choreographic Technologies, and Notational Abstractions through a Laban/Bartenieff Lens*. MIT Press, 2023.

Morriss-Kay, Gillian M. "The evolution of human artistic creativity." *Journal of Anatomy* 216, no. 2 (2010): 158–176.

OED (*Oxford English Dictionary*). 2024 Accessed June 26, 2024. https://www.oed.com/

Ogilvie, Sarah. *The Dictionary People: The Unsung Heroes Who Created the Oxford English Dictionary*. Knopf. 2023.

Raindel, Noa, Yuvalal Liron, and Uri Alon. "A study of dramatic action and emotion using a systematic scan of stick figure configurations." *Frontiers in Physics* 9 (2021): 664948.

Rispoli, Mandy Jenkins, Jessica H. Franco, Larah van der Meer, Russell Lang, and Síglia Pimentel Höher Camargo. "The use of speech generating devices in communication interventions for individuals with developmental disabilities: A review of the literature." *Developmental Neurorehabilitation* 13, no. 4 (2010): 276–293.

Salaris, Paolo, Naoko Abe, and Jean-Paul Laumond. "Robot choreography: The use of the kinetography Laban system to notate robot action and motion." *IEEE Robotics and Automation Magazine* 24, no. 3 (2017): 30–40.

Schoellig, Angela P., Hallie Siegel, Federico Augugliaro, and Raffaello D'Andrea. "So you think you can dance? Rhythmic flight performances with quadrocopters." In Editors: Amy LaViers and Magnus Egerstedt, *Controls and Art: Inquiries at the Intersection of the Subjective and the Objective* (2014): 73–105.

Sher, A. Jang, Umer Huzaifa, Jialu Li, Varun Jain, Alexander Zurawski, and Amy LaViers. "An embodied, platform-invariant architecture for connecting high-level spatial commands to platform articulation." *Robotics and Autonomous Systems* 119 (2019): 263–277.

10

QUESTIONS OF VOICE IN AI MUSIC

Denis Crowdy and James Leach

Introduction

Matters of authorship, ownership, and notions of copyright have become an important focus as AI processes become more prominent in musical activity. This is unsurprising. Machines can potentially generate and/or adapt music to appeal to consumers. This undermines existing models for reward and recognition by human artists. It also raises existential philosophical questions about who or what is creative (in art), and who or what deserves reward for this creativity. These questions are not new and the answers have been far from stable through history, in law, and across cultures.

The relation between existing models for ownership and such questions lies in the assumptions about voice and person that inform those models. The pervasive assumptions about the "natural" rights of authors and creators in Western legal systems are only one answer as to how to think about who creates and who is responsible for communicating art.[1] AI voice generation reveals dilemmas about identity and responsibility as well as reward. Here we outline the kind of concerns that AI voices are raising, locate them within a literature about the connections between identity, a person, and their voice, and then point to resources in the ethnomusicological and anthropological literature for understanding these developments.

Economic imperatives lie behind the motives of authors and artists protesting against the rise of AI tools, but there are also moral concerns involved. It is not only loss of income or trespass on the property of the

DOI: 10.4324/9781003475972-12

creator that is protected by copyright, but a sense of their integrity as a person/author. Copyright includes "moral rights" provisions that protect the author's intentions and reputation. It seems straightforward then to turn to copyright where AI is concerned. The disassociation of a person and their artificially generated or copied voice raises moral and legal concerns that are magnified by fear of the consequences of putting so much of what defines us as human beings in the power of logic machines.

Perhaps the most vociferous moral panic with AI is over deepfakes. Concerns over the destabilizing potential of AI in regard to its apparent or putative ability to reliably mimic and represent a human subject have led to an organised campaign in the US proposing a draft bill to Congress called the "No AI Fraud Act." Known as the "Human Artistry Campaign" (HAC), its proponents seek to "provide for individual property rights in likeness and voice." A summary document and draft of the bill itself exhibits a number of assumptions and statements that open some interesting avenues for comparative analysis. They bring out some of the fundamental questions around ownership, vocal embodiment, and identity that lie behind a wider and more pervasive fear of AI (particularly in Western countries).[2] At the heart of the bill is the statement that "Americans deserve common sense rules to ensure that a person's voice and likeness cannot be exploited without their permission."[3]

It is important to step outside conventional and well-worn reactions to the potential and the threat of AI technologies in relation to music to gain some much-needed perspective. In this chapter we describe theoretical ideas from ethnomusicology and anthropology that summarize what voices are considered to be in different contexts, and how to examine the politics of the circulation of sound through different media. We use these ideas to examine some of the key assumptions underlying the HAC and associated proposed legislation and argue that the perceived threat, and the proposed response, reveal an existential moment for the voice. This takes place alongside more familiar transformations around music as a commodity and new challenges to the effectiveness (or otherwise) of copyright in a largely digital environment.

We first elaborate on the claims of the HAC. We then consider the issues and approaches raised with relevant ethnomusicological theory focused on the idea of acoustemology, an exploration of how we know the world through sound, and schizophonic mimesis, a term that describes the separation of sound from source in the form of recordings, and the politics of the circulation of these recordings. We argue that AI voices offer a particular "schizophonic" juncture with existential implications in Western societies. We ask questions about AI vocality, identity, indexicality, and what histories and possibilities emerge from the circulation of the resemblance of another's voice.

In conclusion, we suggest that current reactions attempting to extend IP laws do not do justice to the complexities involved in these rapidly developing AI music scenarios. Turning to robotics, we show that it is worthwhile to consider the kind of re-embodiment involved in the production of recognisable voices by machines, and by extension the kinds of bodies these might be coming from, alongside the matters of identity and ownership that accompany such re-embodiment.

The Human Artistry Campaign

On loading, the HAC campaign website offers a popup to support the No AI Fraud Act with the motto "It's your voice – you should be the one to decide what it says." This sounds so obvious as to be unarguable. However, the assumption that one can own or does own a voice is at least worth considering carefully. Removing the popup, the main page carries the heading, "Creative works shape our identity, values, and worldview" with the word "creative" circled as though having been hand-drawn by a person. While the statement is straightforward, the circling of "creative" in this manner intimates a questioning of whether machines can be creative. This emphasis on creativity is important to reinforce the contrast about to be set up between human and machine creativity. Indeed, machine creativity is denied by the exclusivity expressed in the next paragraph on the page:

> And there are fundamental elements of our culture that are uniquely human. Only humans are capable of communicating the endless intricacies, nuances, and complications of the human condition through art – whether it be music, performance, writing, or any other form of creativity.

The idea of there being fundamental elements that are uniquely human seems reasonable, but the step to the idea that only humans are capable of communicating the human condition through art has been questionable for some time (Dahlstedt 2009, 2021). Clearly, in the case of deepfake musical examples provided in the draft Act, people *have* valued creations where a machine has generated a recognisable human voice. The one-page summary of the draft Act provides the following examples:

> From an AI-generated Drake/The Weeknd duet, to Johnny Cash singing "Barbie Girl," to "new" songs by Bad Bunny that he never recorded to a false dental plan endorsement featuring Tom Hanks, unscrupulous businesses and individuals are hijacking professionals' voices and

images, undermining the legitimate works and aspirations of essential contributors to American culture and commerce.

One of the problems the HAC faces is that although they see this as activity of unscrupulous hijackers, in some cases music has been created that is valued by a significant number of people. The introduction of the bill recognizes this when it includes statistics like these:

> In fall 2023, AI technology was used to create the song titled "Demo #5: nostalgia," manipulating the voices of Justin Bieber, Daddy Yankee and Bad Bunny. It reportedly received 22 million views on Tok Tok and 1.2 million views on YouTube.

Putting aside for the moment whether it was their actual voices that were manipulated, the campaign is clearly concerned about the effects of this music's circulation and the challenges it presents to established ideas of ownership and value.

The HAC recognizes there is value in AI technologies in a paragraph that suggests a certain inevitability while drawing a line about human creativity:

> Developments in artificial intelligence are exciting and could advance the world farther than we ever thought possible. But AI can never replace human expression and artistry.
>
> As new technologies emerge and enter such central aspects of our existence, it must be done responsibly and with respect for the irreplaceable artists, performers, and creatives who have shaped our history and will chart the next chapters of human experience.

Artist intention and agency is the value behind this; so, for example, it might be fine for The Beatles to release a new song using AI (as they have done without widespread complaint), but someone else generating the effect of the late Johnny Cash singing "Barbie Girl" without his estate's permission might not. And finally:

> These principles outline how we can responsibly use artificial intelligence – to support human creativity and accomplishment with respect to the inimitable value of human artistry and expression.

The use of "inimitable" is at the center of a significant problem of circularity here. Expression through art that has previously been the domain of humans is being carried out by machines. What if machine-generated art has significance for at least some humans? What if some people are unable to differentiate between art created by humans and machines – a kind of artistic Turing test? If those questions are answered in the positive, then the idea of inimitability in some universal human sense is untenable. What then is the complaint based upon? What is it then that a voice consists of?

Ownership of the voice and authorship by machines powered by AI emerge as key concerns from the HAC without much clarity. There is a commercial imperative in the attempt to protect artists who have built a career and earn a living based on the sound of their voice and what it sings, and when. But it is more than the commercial loss that is being highlighted. Humanity and individuality seem at stake.

We move to explore issues of ownership of the voice and associated ontological questions. We reference existing work that has examined the politics of circulating sound that indexes people and places where those sounds have been separated from their sources by being in a recorded format. Applying this work to AI-generated voices, we argue, exposes an existential moment for the voice, both as a recognisable sonic phenomenon and as a medium for speaking with implications for individuality, authenticity, and authority. This discussion will be prefaced by an excursion into ethnomusicological perspectives on copyright. One significant aspect of copyright is that it has always protected expressions, not ideas. That is to say, created objects, not the capacity or ability to create them. We then discuss how these particular examples fit into similar disciplinary analyzes around circulation, value, and commodity status.

Ethnomusicology on Copyright

A typical Global North/Euro-American perspective on challenges to ownership and authorship is to adapt copyright law, and in this case, extend the reach of intellectual property law to the voice. To complicate matters, in the HAC discourse, there are several references to human universals. That comes undone in relation to copyright, as despite the widespread uptake of copyright internationally through agreements such as Trade-Related Aspects of Intellectual Property Rights (TRIPS) and advocacy from the World Intellectual Property Organization (WIPO), there is huge variation in the effectiveness of copyright for music globally. Indeed, as Lysloff points out, the idea is meaningless for many:

> technologies are used by a people in ways that are appropriate to that particular society—that is, technologies often have different meanings and values in different cultural contexts. This is exemplified by the way

> audio and video recording technologies have given rise to third-world piracy. The argument that pirating music recordings or computer software is morally wrong, implicating people who could not possibly afford a legitimate copy of, say, Microsoft Windows, simply has no meaning for most Indonesians. Indeed, copyright enforcement is viewed as an unfair way to deny poorer nations access to knowledge and entertainment.
>
> *(Lysloff 2016, p 487)*

Leach and Stern (2020) point out some of the complexities, inequities, and inconsistencies in the uptake, application, and infrastructure for music copyright in Melanesia. Even in places where copyright is entrenched with effective infrastructure, there are considerable differences in how it is applied (between US/UK/Europe/Canada/Australia, for example). There are important differences even between legal systems that are closely related historically (the UK and Australia). Copyright may seem a simple and natural way to protect the voice of a person. However, these inconsistencies show the particularity of the formulation of the law, which is focused on recognizing individual creators and assigning their rights through the medium of property-over-commodities produced. It does not and cannot cover the unique processes of creation artists might have at hand. Similarly, moral rights protect created expressions, not the ability to create, or the medium/instrument of expression.

Circulation, Exchange, and Value

The ability of many more people to be able to manipulate a voice and then distribute the results widely is a key component of the perceived threat the HAC is responding to. The important element of this for our analysis is circulation. The kind of object-in-circulation that music is, has been, and might be is significant here when considering arguments made by the HAC that are largely focused on economic harm while appealing to a more amorphous sense of morality and human uniqueness. Music is a complicated object. It is not stable as a commodity, as through its "life" its status shifts between various forms of which commodity is just one. Its circulation under different conditions generates different kinds of value. As a result, and because of the flexibility in how music is valued and used (McDonagh 2022), there is the possibility for use value to overwhelm exchange value. Indeed, in some ways, it is the "humanity" of use value – that is, music and the voice's role in creation and innovation, and in social solidarity – that gives moral force to the economic concern.

> I would suggest that through its latent subversive potential, music is only ambivalently a commodity. Its commodity status must be enforced, and is resisted, at all turns by the people and institutions that seek to keep it in its commodity status as long as possible. Only with rigorous policing, lobbying, legislation, and enforcement does music fit, however reluctantly, into the industrial mould that has been cast for it.
>
> *(Beaster-Jones 2014, 337)*

Ethnomusicology, with its integral relationship to anthropology, has long been concerned with the value of music for social function and its role in social reproduction. AI can be considered just another kind of sound-reproducing technology, and these technologies always sit within specific social relations of production and circulation. Technology has and tends to continue to lead the law. That has certainly been the case with sound-reproducing technology. With recorded music, we have a situation where technology has periodically challenged music's commodity status – stretching back to the pianola, radio, LPs, tapes, mp3 files, streaming, and so on. One of the most illustrative cases is that of Sony in the 1990s, simultaneously developing DAT recorders with their ability to clone recordings that were being made elsewhere in the company as a major music label. We say clone because the difference from a copy is important, and this difference is a direct result of the change that the new technology enabled. If copy implies a not-quite-perfect mimesis, then digital copies are clones, in that they are exact replicas. There is no difference between original and copy, and thus they cannot be distinguished. Indeed, the idea of "original" subsides in this situation. This has seemingly profound implications when it comes to human identity because of the strong association of voice with individual people. The kind of mimesis involved in AI and deepfake voices makes this clear. But importantly, it is not just a clone we are hearing here, but something else as well. We are dealing with a clone that indexes human intention and agency. We explore this in the following section.

Machine-based Vocal Schizophonic Mimesis

We know the world in important ways through the voice; through speaking, singing, and listening. Voices are ways of sounding and simultaneously creating and interacting in a sonic and cultural milieu. Steven Feld has coined the term "acoustemology" – a portmanteau of "acoustic" and "epistemology" – to frame thinking about the importance of listening in general to "foreground sonic experience as a way of knowing" (Rice 2018, 1). Greene and Porcello (2005) offer a modified term – techoustemology – "to foreground the implication of forms of technological mediation on individuals' knowledge and interpretations of, sensations in, and consequent actions upon their acoustic environments as grounded in the specific times and places of the production and reception of sound" (p270). While that term might be more useful for the AI case, the main idea of both is that our sense of hearing and environment are linked in important ways that define senses of place, identity, and relationships.[4] Studies on how we come to know the world through listening have brought our attention to two fundamental aspects of voices. These are firstly embodiment, and secondly, voice as an aspect of identity, politics, and social position. Voice is not just an object of art, but integrated into how we know ourselves and others and how we reveal intention and agency.

Feld et al. have summarised studies in vocal anthropology as revealing the "micropolitics of emplaced, embodied, and voiced identity in particular local lifeworlds" (Feld et al., 2005, p. 341). Embodiment creates a connection with how we recognize and understand people and places. The voice indexes people in particular places where their bodies are located:

> it is always the body social that is enunciated in and through the voice.... the sonorousness of the voice indexes a clear social agency and a sense of place.
>
> *(Feld et al., 2005, p. 341)*

That vocal indexicality is a fundamental part of how we construct, continue, and develop our knowledge of the world:

> Voice is the embodied locus of spoken and sung performance, the site where language and music have received closest ethnographic scrutiny. [...] Vocality, in this light, is a social practice that is everywhere locally understood as an implicit index of authority, evidence, and experiential truth.
>
> *(Feld et al., 2005, p. 341)*

Indexicality has been challenged by sound in the past through the creation and circulation of audio recordings and a key way of thinking about this has been through the notion of schizophonia. AI voices are a kind of audio recording, but not of the person being imitated. That is really the point of deepfake, of course. Schizophonia is a term coined by R. Murray to describe the effects of breaking indexical links between source and sound (Schafer 1969). Feld developed the idea further, particularly in relation to the mimetic use and re-use of ethnomusicological recordings in commercial "world music" contexts (Feld 1994, 1996). This passage is worth quoting at length for its significance in application to AI voices:

> I coin the phrase "schizophonic mimesis" here to point to a broad spectrum of interactive and extractive practices. These acts and events produce a traffic in new creations and relationships through the use, circulation, and absorption of sound recordings. By "schizophonic mimesis" I want to question how sonic copies, echoes, resonances, traces, memories, resemblances, imitations, duplications all proliferate histories and possibilities. This is to ask how sound recordings, split from their source through the chain of audio production, circulation, and consumption ... stimulate and license renegotiations of identity. The recordings of course retain a certain indexical relationship to the place and people they both contain and circulate. At the same time their material and commodity conditions create new possibilities whereby a place and people can be recontextualized, rematerialized, and thus thoroughly reinvented. The question of how recordings open these possibilities in new, different or overlapping ways to face-to-face musical contacts, or to other historically prior or contiguous mediations remains both undertheorized and contentious.
>
> *(Feld 1996, 13)*

A disembodied voice that sounds like a person emanating from a machine exhibiting unpredictable agency challenges the indices linking humans to one another. "Renegotiation" of indexicality and identity is happening in the appropriation. Deepfake AI voices retain an indexical relationship to "the place and people." They reference these people and places and thus identity. When AI voices appear to act in human-like ways, they are both indexing the person's value significance and appropriating that person's potential agency. By associating a machines' creativity with a person's voice, they hide and misrepresent the source. As a result, indexes of authority, evidence, and experiential truth may be manipulated, even severed. The things being said by an AI voice do not need to have any relationship at all to any intentions of the human whose voice has been imitated. The index of identity is weakened in an inverse relationship to the strength in vocal similarity.

The voice is not an object but a unique medium. It is in the act of speaking or singing that objects can be created that do critical parts of any indexing work. An AI vocality re-embodies the voice in a very different kind of body. When that voice bears a likeness to a voice we are used to hearing emanating from a human body, a severing rather than renegotiation of indexicality and identity occurs. It is this particular circumstance that the HAC is railing against. But as we have pointed out, copyright as a version of "ownership," as it stands, cannot be an answer here. Appeals that assume a natural ownership of the voice in terms of property are likely to fail.

Alternate Bodies

Further to thinking through the implications of non-human voices, and of the detachment of voice from person, there are numerous examples describing non-human voices in the anthropological literature. These are often closely associated with people, but provide examples in which speaking through or with another voice holds different connotations. In our regional area of expertise of Papua New Guinea (PNG), for example, numerous ethnographic studies describe this in great detail. Yamada (1997) has studied the sound world of the Waxei people from the East Sepik Province of PNG and details how voices of spirits emerge through performances on bamboo flutes. This is a wider spread phenomenon, as Niles (1992) has described through the use of voice-distorting instruments. Such examples illustrate the different ways that people, places, and voices may be heard, recognised, and come to have significance for identity. They reveal how the position and significance of voice rely on different assumptions about the relation of voice to a person. The analytic approach outlined in acoustemology and schizophonia is valid across domains to discover these relations.

There are also relevant ethnographic cases of established ways in which a person's public voice is considered a dialogic or collaborative achievement. It is thus not "owned" by one person and potentially indexes several people or specific groups. Slit-gongs, or *garamuts* in Tok Pisin, have served as instruments for long-distance communication (Watson and Duffield 2016), and their production and use are tied to important ancestral traditions and processes. Slit gong-drums made by Nekgini speaking people in Madang Province of PNG, for example, are the audible presence of the men they are constructed for (Leach 2002). But a man does not make his own drum. It is made by his wife's family and given to him as a final recognition of successfully completing the payments required for marriage. His ability to "speak" in this example – his voice – is strongly associated with a relationship and multiple people. It is only "his" voice because it already contains, as it were, multiple people and their intentions and agency. While it is generated as an aspect of an identity that is achieved, that achievement is a collaboration, and is not his property to dispose of as he wishes. In this case, firstly, the personal voice is given power and presence because of its detachment from the person themselves. Secondly, responsibility for vocality is a collective responsibility and the ability to project agency through a voice is socially constituted. Thirdly, different modes of ownership result. There is a great deal more to explore here.

Through such examples, we become aware of alternative historical and cultural connections to people and places through the voice. Making the shift from geographically bound instances to the plethora of possibilities through digital distribution can be done by thinking about other socially constituted voices that have become important in recent times. The presence of synthetic voices such as Siri and Alexa as AI voice assistants emerge from a complex socially and industrially constituted set of processes. These are dynamic in that the AI engines they draw on and represent are constantly in flux as they create identities and generate assumptions and expectations as they are used. Holliday (2023) has examined how people have interpreted different offerings of Siri's voice, bringing into question gender and cultural background and Woods (2018) explores the politics of vocally encoding "gendered characteristics to pacify anxieties about embedding surveillant devices within the intimate sphere" (p. 346).

Robot Voices

If the HAC's concern is with the replication of an index of human agency without responsibility, then developments in robotics offer a different approach. In fact, one of the most important places for machine voices is the field of robotics. Communication between humans and robots is a significant topic of interest to people designing and building robots. The field of Human Robot Interaction (HRI) has examined how robots might best use sound in a variety of forms – voices – to interact with people more effectively. The idea of robots and their voice being robotically embodied is a core idea with a long history. Some researchers in this field, for example, use this idea as a base from which to consider the implications of being technically able to disembody sound from a robot:

> From a human–robot interaction (HRI) perspective, robot sound being able to leave a robot's body has an additional implication. Using an anthropomorphism, the robot may be considered a body (hardware) and a mind (software) that can be disembodied. While many of the robot mind's capabilities are shared with other entities like voice assistants, the robot's physical body is a distinctive characteristic.
>
> *(Robinson, Velonaki, and Bown 2024, 129)*

These minds-as-software use language through a voice in a mechanical body and interact with the world. So much so that the idea of designing robots that follow social norms, sonic ones included, has been central (Krantz, Balkenius, and Johansson 2024, 62).

Often missing from discussions on AI and music is the extent to which machines have, for quite some time, been imagined and represented with voices (The Daleks in Doctor Who, Hal the computer in the film 2001). In music, Radiohead (among others) have incorporated voices we associate with Apple computers (*Fitter Happier* from the 1997 album OK Computer, for example), and vocoders and audio artifacts from autotune effects have rendered human voices less individually recognisable and more machine like. There are a variety of contexts where recognisable machine voices are highly valued for the work they perform, from screen-readers for sight-impaired people to AI personalities in various relationships with humans.

Embodiment has been conceptualised in a different manner here. An embodied voice is valued in robotics as an interface between humans and machines. The robot "owns" its own voice. AI deepfakes hide the machine.

Conclusion

Technology development and cross-cultural and historical analysis both make apparent some narrow conceptualisations underlying concerns about AI voice, while copyright laws are often formulated through lobbying to protect powerful commercial interests. They are not concerned with use value or personal identity.

We value machine voices and incorporate them into human music culture. Looping back to Feld, what remains under-theorised and contentious is how identity, and what kind of identity, emerges as these voices sing and speak. If embodiment has always been a key part of our understanding of the voice, and the act of disembodiment lies at the core of challenges to identity and representation, then as long as we recognise that we do value machine voices in particular situations, we perhaps need to turn to think about what sort of re-embodiment is entailed by AI voices. Ethnomusicology and anthropology have important contributions to make in this regard.

What is important is developments in AI tools have given people the ability to manipulate something previously not possible (an identifiable voice). Although an existential crisis like this demands some sort of response in the case of deepfakes, the idea that "Every individual has a property right in their own likeness and voice" is in fact an ambition and not a statement of fact. How one would make that a reality is unclear as the law of copyright on which the music industry has been reliant seems to offer limited options. These limitations are a result of principles about the person and property that are specific to recent Western society. One thing our exploration has detailed is that a voice, an idea of embodied vocalised identity, cannot simply be afforded the status of intellectual property. It is a medium of expression and not the expression. The controversy around AI voice illustrates the limitations of this distinction. Where the medium is also personal and an aspect of embodiment, property is a poor tool with which to capture its value.

Notes

1 For example, see Bowden, Ross 2022 *Art and Creativity in a New Guinea Society. The Kwoma in a cross-cultural perspective.* Lanham: Lexington Books, 2022.
2 See, for example, ABC News (2023, 2024), BBC News (2023).
3 By signaling "common sense," the authors thereby appeal to naturalised or assumed principles of the person in modern US society. We are not going to deal with aspects of visual likeness but suggest that Image Rights (Australia) or Publicity Rights (US) may be a promising avenue for further exploration here, as the same issues that we have identified about the voice, including indexicality, and circulation as social value creation will be relevant to their use.
4 Feld and Basso 1996.

References

ABC News. 2023. "'The Law Is Light-Years behind': The Battle over Training AI on Copyrighted Work," November 28, 2023. https://www.abc.net.au/news/science/2023-11-29/artificial-intelligence-ai-training-datasets-copyright-books3/103157980

ABC News. 2024. "Future Looks Bleak for Voice Artists Losing Work to Their AI Clones," May 23, 2024. https://www.abc.net.au/news/2024-05-24/australian-voice-artists-losing-work-to-their-ai-clones/103885430

BBC News. 2023. "AI-Generated Drake and The Weeknd Song Goes Viral," April 17, 2023, sec. Entertainment & Arts. https://www.bbc.com/news/entertainment-arts-65298834

Beaster-Jones, Jayson. 2014. "Beyond Musical Exceptionalism: Music, Value, and Ethnomusicology." *Ethnomusicology* 58 (2): 334–40. 10.5406/ethnomusicology.58.2.0334

Dahlstedt, Palle. 2009. "Thoughts on Creative Evolution: A Meta-Generative Approach to Composition." *Contemporary Music Review*, February. 10.1080/07494460802664023

Dahlstedt, Palle. 2021. "Musicking with Algorithms: Thoughts on Artificial Intelligence, Creativity, and Agency." In *Handbook of Artificial Intelligence for Music: Foundations, Advanced Approaches, and Developments for Creativity*, edited by Eduardo Reck Miranda, 873–914. Cham: Springer International Publishing. 10.1007/978-3-030-72116-9_31

Feld, Steven. 1994. "From Schizophonia to Schismogenesis: On the Discourses and Commodification Practices of 'World Music' and 'World Beat'." In *Music Grooves*, edited by Charles Keil and Steven Feld, 257–89. Chicago: The University of Chicago Press.

Feld, Steven. 1996. "Pygmy POP. A Genealogy of Schizophonic Mimesis." *Yearbook for Traditional Music* 28 (January):1–35. 10.2307/767805

Feld, Steven, Aaron A. Fox, Thomas Porcello, and David Samuels. 2005. "Vocal Anthropology." In *A Companion to Linguistic Anthropology*, 321–45. John Wiley & Sons, Ltd. 10.1002/9780470996522.ch14

Greene, Paul D., and Thomas Porcello, eds. 2005. *Wired for Sound: Engineering and Technologies in Sonic Cultures*. Middletown, CT: Wesleyan University Press.

Holliday, Nicole. 2023. "Siri, You've Changed! Acoustic Properties and Racialized Judgments of Voice Assistants." *Frontiers in Communication* 8 (April):1116955. 10.3389/fcomm.2023.1116955

Krantz, Amandus, Christian Balkenius, and Birger Johansson. 2024. "Talk to Me: Using Speech for Loss-of-Trust Mitigation in Social Robots." In *Sound and Robotics*, edited by Richard Savery, 1st ed., 1:61–75. New York: CRC Press.

Leach, James. 2002. "Drum and Voice: Aesthetics and Social Process on the Rai Coast of Papua New Guinea." *Journal of the Royal Anthropological Institute* 8 (4): 713–34. 10.1111/1467-9655.00130

Leach, James, and Monika Stern. 2020. "The Value of Music in Melanesia: Creation, Circulation, and Transmission Under Changing Economic and Intellectual Property Conditions." In *The Oxford Handbook of Economic Ethnomusicology*, edited by Anna Morcom and Timothy D. Taylor, 1st ed. Oxford University Press. 10.1093/oxfordhb/9780190859633.013.33

Lysloff, René T. A. 2016. "Worlding Music in Jogjakarta: Tales of the Global Postmodern." *Ethnomusicology* 60 (3): 484–507. 10.5406/ethnomusicology.60.3.0484

McDonagh, Luke. 2022 "Exploring 'Ownership' of Irish Traditional Dance Music: Heritage or Property? | *International Journal of Cultural Property*." Cambridge Core. Accessed February 4, 2025. 10.1017/S094073912200011X

Niles, Don. 1992. "Konggap, Kap and Tambaran: Music of the Yupno/Nankina Area in Relation to Neighbouring Groups." In *Abscheid von Der Vergangenheit*, edited by Jurg Wasserman, 149–83. Berlin: Deitrich Reimer.

Rice, Tom. 2018. "Acoustemology." In *The International Encyclopedia of Anthropology*, 1–7. John Wiley & Sons, Ltd. 10.1002/9781118924396.wbiea2000

Robinson, Frederic Anthony, Mari Velonaki, and Oliver Bown. 2024. "Robot Sound in Distributed Audio Environments." In *Sound and Robotics*, edited by Richard Savery, 1st ed., 1:128–58. New York: CRC Press.

Schafer, R. Murray. 1969. *The New Soundscape: A Handbook for the Modern Music Teacher*. Don Mills, Ont: BMI Canada.

Watson, Amanda H A, and Lee R Duffield. 2016. "From Garamut to Mobile Phone: Communication Change in Rural Papua New Guinea." *Mobile Media & Communication* 4 (2): 270–87. 10.1177/2050157915622658

Woods, Heather Suzanne. 2018. "Asking More of Siri and Alexa: Feminine Persona in Service of Surveillance Capitalism." *Critical Studies in Media Communication* 35 (4): 334–49. 10.1080/15295036.2018.1488082

Yamada, Yoichi. 1997. *Songs of Spirits: An Ethnography of Sounds in a Papua New Guinea Society*. Vol. 5. Apwitihire: Studies in Papua New Guinea Musics. Boroko [Port Moresby]: Institute of Papua New Guinea Studies.

11

WHAT DO I SAY? PUBLIC INTERACTIONS WITH A DRUMMING AND RAPPING ROBOT

Richard Savery, Trinity Melder and Melissa Hill

1 Introduction

The integration of robots into public spaces has been an evolving phenomenon, with their roles expanding from functional tasks to more interactive and social dimensions (Seo and J. H. Lee, 2021; Wilkinson, Bultitude, and Dawson, 2011). In these settings, robots are not only tools but entities that engage with the public in various forms, from providing information and assistance to entertaining (Foster et al., 2016). One developing aspect of this interaction is the communication between humans and robots, particularly what people say and how they say it. This form of interaction is rich in nuances and offers insights into human-robot dynamics in public settings.

Robotic musicianship is an emerging interdisciplinary field combining music, computer science, performance studies, AI, and mechatronics (the physical creation of robotics) (Frid, 2023). The field of robotic musicianship aims to develop robots with musical abilities, such as to listen and improvise to human performers (Savery, Zahray, and Weinberg, 2021b). Music styles explored in robotic musicianship, including jazz improvisation and free style hip hop interaction (Savery, Zahray, and Weinberg, 2020), offer communicative dialogue-based mediums that require extensive domain knowledge. Music, as one of the most culturally significant, social and personal experiences, offers powerful opportunities to investigate the challenges of incorporating advanced innovative technologies into new domains and society.

In this work, we utilise Keirzo, a custom-built robot with capabilities in drumming and rapping as our medium to study robot performance.

DOI: 10.4324/9781003475972-13

Keirzo is specifically designed to engage with people in a public performance, as shown in Figure 11.1. This robot is equipped with the ability to comprehend human speech and respond in a musically creative manner, thus serving as an ideal subject for our investigation into the dynamics of human-robot interaction in an informal but busy public environment. When interacting with Keirzo, the audience themselves join the performance, with their voice amplified through the system.

We developed an exploratory research approach to analyse the dynamics of human-robot interaction (HRI) within the environment of a public festival. Our investigation focused on developing understanding of the nature of communication between humans and robots in such a setting, where spontaneous and unscripted encounters provide a wide range of verbal exchanges. By focusing on these interactions, we aimed to uncover not just the content of what people say to a robot but also the underlying sentiments, questions, and requests that emerge in the presence of robotic performance technology and how music shapes these interactions. This approach allowed us to capture a snapshot of public perceptions and

FIGURE 11.1 Keirzo: drumming and rapping robot. (credit Richard Savery).

attitudes towards robots, providing insights into how people of varying ages and backgrounds engage with and react to the presence of a robot among them.

A key objective of our research was to differentiate between the performance interactions initiated by children (under 13) and those by adults, recognising that age plays a significant role in shaping the nature and content of human-robot communication and all social interaction. By comparing these interactions, we sought to identify distinctive patterns, themes, and levels of curiosity or scepticism that might exist between these two demographic groups. Furthermore, we aimed to gauge the general attitudes towards artificial intelligence and robotics, exploring whether the interactions reveal broader societal views, concerns, or expectations about the integration of such technologies into everyday life. Through this exploration, we hoped to contribute to the ongoing discourse on the social acceptance of robots, providing valuable data on how different segments of the population perceive and interact with these increasingly pervasive technological entities.

Our research was conducted at the Parramatta Lanes festival in 2023, where Keirzo interacted with a diverse public population (Figure 11.2). Our methodology involved recording and analysing the data from these interactions, using a combination of natural language processing and manual categorisation to dissect the nature of the dialogues and responses. This approach allowed for a comprehensive understanding of the interaction patterns, taking into account both the content and context of the communications. Our findings reveal multiple aspects of how people of different ages and backgrounds engage with a robotic entity in a creative and public setting. The study uncovers patterns in the types of queries posed, the nature of the requests made, and the overall sentiment of the interactions when audience members interact with a robotic musician.

2 Related Work

Recent studies have documented the expanding roles of robots beyond industrial applications, venturing into public spaces where they interact directly with the general populace. However, the novelty of encountering robots in such settings often elicits mixed reactions from the public, shaped largely by prior exposure to technology and cultural representations of robots (Sumartojo et al., 2021; Tian et al., 2020). For instance, multiple researchers discuss how cultural portrayals influence public perceptions, with varying degrees of scepticism or acceptance observed across different societies (Bartneck et al., 2020;

Sone, 2017). These studies underline the importance of context in shaping the public's reception of robots, suggesting that design and functionality should be clearly communicated to mitigate apprehensions and enhance engagement.

The interaction between humans and robots or AI has evolved, with conversational agents like ChatGPT becoming increasingly commonplace in daily life. While these interactions demonstrate the potential for communication with AI, they also reveal a spectrum of human engagement styles, from informational queries to seeking emotional connections (Gao, Galley, and Li, 2018). Other research provides insights into how the nature of initial human approaches – whether seeking information or testing the AI's capabilities – affects the subsequent interaction style (Raza et al., 2022). Similarly, past research explores the impact of different robot speech patterns and the willingness of humans to engage in deeper, more personal conversations with robots (Pan et al., 2015;). Other research has explored the potential of emotion-infused communication when interacting with individual robots (Savery, Rogel, and Weinberg, 2021a) or groups of robots (Savery, Zahray, and Weinberg, 2021c) for improved trust and likeability. These findings suggest that not only do HRIs mirror the complexity of human-human communication, but they also reflect broader societal attitudes towards technology.

Despite these advancements, a coherent understanding of how different demographic groups, particularly children versus adults, navigate these interactions remains underexplored. Moreover, there is a lack of clarity on how these interactions reflect broader societal views on the integration of AI and robotics into daily life. This gap highlights the need for research that systematically examines the nature of communication between humans and robots in public settings, taking into account demographic differences and societal attitudes and the impact of creative robotics on these cultural aspects.

3 Method

This study adopts an exploratory research approach to the nuances of HRI in public spaces. The aim is to uncover the various dimensions of how people engage with robotic technology, focusing on verbal communications, comparative analysis between adults and children, and the broader societal implications these interactions may hold. Exploratory research is particularly suited for this study as it allows for the discovery of insights into complex behaviours and attitudes that have not been extensively quantified or fully understood in the context of public robotics. Our goal

FIGURE 11.2 Keirzo at the festival. (credit Richard Savery).

was to facilitate real-world interactions – as opposed to Wizard of Oz scenarios, which are pre-scripted in a research lab – to understand how autonomous robotic engagements function in public settings.

3.1 Research Questions

RQ1: What characterises the verbal interactions between the general public and a musical performance robot?

This research question seeks to catalogue and analyse the specific content of verbal interactions initiated by the public with the robot. This will begin by examining whether participants direct questions, comments, or requests at the robot. This question aims to provide a foundational understanding of direct public engagement with robotic technology in a communal setting. By focusing on the immediate and unfiltered responses of people in such settings, this question addresses the core elements of HRI, including the types of verbal exchanges, the level of curiosity and engagement, and the social norms exhibited during these interactions. Understanding these dynamics can reveal underlying patterns and trends in public perception and acceptance of robotic technology, thereby contributing to the broader discourse on the social integration of robots.

RQ2: What are the key differences in the nature of questions and requests made to a musical performance robot by children compared to adults?

Focusing on a comparative analysis, this question investigates how interactions with the robot differ between children and adults. It explores the variance in curiosity, the complexity of inquiries, and the nature of requests, offering insights into age-related differences in human-robot communication.

RQ3: In what ways do the interactions with a musical performance robot reveal participants' perceptions, fears, and expectations regarding the future of robotics and AI technology?

This question explores the underlying attitudes and emotions that inform public interactions with the robot. By analysing the content and sentiment of these interactions, it aims to uncover broader societal views on the integration of robotics and artificial intelligence into everyday life.

3.2 Location and Demographics

Parramatta Lanes is an annual street festival, that takes place in the Parramatta, Sydney, Australia. In 2023, the year of this study, the event was attended by 180,000 people. The event is attended primarily by locals to the area, as well as audience members from the wider Sydney area. The event is renowned for showcasing a variety of local cuisines and features an array of street performers and installations, creating a vibrant and culturally rich atmosphere. Parramatta is ethnically diverse, with 40% of the population born outside of Australia, and 12% identifying their ancestry as Australian, with the top five self-identified ancestries as Indian, Chinese, English, Lebanese, and Filipino (Australian Bureau of Statistics, 2021).

The festival took place from October 11 to October 14, operating each day from 5:00 pm to 11:00 pm. Throughout the duration of the festival, the robot was active and engaging with attendees.

Our robot, referred to in the programme as "Meet the Robot Keirzo," was a highlighted feature of the festival. The authors of this study were present on-site throughout the event, observing and recording the interactions between the public and the robot. The research process, including interaction protocols and data collection methods, adhered to ethical guidelines and received approval from Macquarie University Ethics.

3.3 Interaction Protocol

Our study used Keirzo, a custom robot platform designed with a focus on HRI. Keirzo's design incorporates focuses on engaging humans, featuring multiple subsystems that allow for both manual programming and autonomous operation. In autonomous mode, Keirzo can track musical beats and respond to spoken audio cues. Key features include a head with three degrees of freedom, an integrated digital screen, a "rib cage" simulating breathing, circular LEDs, four drumsticks for musical performance, and two speaker outputs. For the festival performance, the motorised wheels were omitted to ensure safety and manoeuvrability in the crowded environment.

Keirzo's interaction capabilities are powered by a pipeline integrating speech recognition, generative AI, and a unique audio output system (Savery & Savery, 2024). Using the Whisper API (Radford et al., 2023) for speech-to-text conversion, Keirzo processes spoken words from the audience and generates rap responses through a generative AI model. This AI-driven text generation and rap synthesis, detailed in a separate publication by Savery, Zahray, and Weinberg (2020), are complemented by algorithmic drum grooves that blend with Keirzo's rapping.

A microphone was set up in front of Keirzo to facilitate direct communication with participants. A poster labelled "Talk to Me" was placed near Keirzo to prompt audience interaction. Data collection for our study was multi-modal, including auditory responses from participants transcribed into text and stored directly in the robot's system, as well as video recordings for context and clarity. All data collected during Keirzo's performance at the event were included in our study.

4 Results

Our analysis of the recorded data focused on combining basic natural language processing with human-generated labels for the interactions. In this section we first present a breakdown of participant timing and percentage of adults or children. We then show our categorisation process, with broad categories manually derived and then followed by subcategories from these sections. Finally, we explored keyword usage separated into children and adults. In our analysis we omitted responses – 27 from children and 26 from adults – that we were unable to understand, due to issues such as distance from the microphone and languages other than English.

4.1 Interaction Times

We had a total of 312 children interact with the robot and 499 adults. We tracked only those who interacted directly with the robot, for example, for children accompanied by a parent we only counted the child, unless the adult also spoke to the robot. The robot was officially open from 17:00 to 22:00; however, participants often chose to speak to the robot as soon as it was being tested and continued interacting after hours. We found no significant variation in the types of conversations that happened over the night at varying times. At the busiest time of day from 19:00 to 20:00 the robot had a total of 232 interactions, averaging 58 per day at that time.

4.2 Broad Categorisation

In our keyword analysis, we first organised the collected data by segregating the responses. We utilised all responses, both from adults and children, while excluding the responses generated by the robot Keirzo. The human responses were then categorised into two distinct groups: those from children and those from adults. This categorisation was facilitated by using video recordings of the interactions, which allowed us to accurately associate each response with the corresponding participant.

We conducted an initial screening of the child and adult datasets to identify and remove any responses that were in foreign languages or difficult to understand. Following this, we further divided the remaining data based on the type of expression used in each response. This subdivision led to a more detailed quantitative analysis of the datasets.

Our research revealed various forms of casual communication and queries. The responses were split into six categories:

- Agreements/Disagreements: Responses agreeing or disagreeing with the robot's past statement.
- Exclamations: Expressions marked by enthusiasm or excitement.
- Greetings: Initial expressions of salutation or acknowledgement at the beginning of interactions.
- Statements: Authoritative or declarative expressions sharing information or opinions.
- Questions: Inquiries seeking information or clarification.
- Requests: Appeals for action, favours, or specific contributions.

We observed that audience members often interacted with Keirzo by asking questions, which constituted 34.84% of all collected responses. This was followed by making requests (15.71%) and extending greetings to

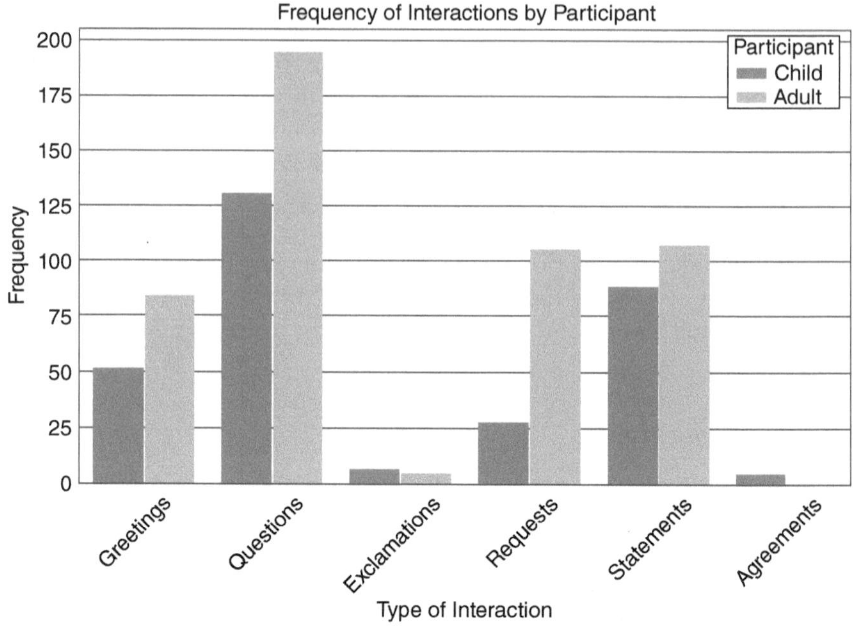

FIGURE 11.3 Frequency of interactions. (credit Richard Savery).

Keirzo (14.74%). The overall frequency of interactions is presented in Figure 11.3, where each category of interaction is tallied.

4.2.1 Introductions and Basic Greetings

In our dataset, we define greetings as any form of basic introduction. To separate greetings from other responses, we used a predetermined set of criteria or keywords to identify and categorise these instances. We identified specific keywords or phrases commonly associated with greetings. For example, words such as "hello," "hi," "hey," or other friendly salutations were used to identify greeting responses. The surrounding context of the responses was also considered. Responses that primarily consisted of friendly and introductory expressions, with a lack of specific questions, requests, or statements, were categorised as greetings.

The percentage breakdown indicates that greetings constituted 43% of children's responses and 57% of adult responses. In 32.69% of instances children used the phrases "Bye" or "Bye Keirzo" in their queries to Keirzo, and 40.38% said "Hi" or "Hi Keirzo." 1.92% of the times children used greetings such as "*Have a great day*" and "*dear.*" 13.46% of the times

"*How are you?*" was used. "*See you*" and "*Good Night*" was used 5.77% and 3.85% times respectively.

Adults used "*Bye*" phrases 13.04% of the times in their greetings and 56.52% times the phrases "*Hello, Hi, Hey, How are you!*" "*Good Afternoon!*" and "*Cheers!*" was used 1.45% of the times and "Okay" was used 7. 25% of the time. Other types of phrases like "*Good playing buddy,*" "*Wish me happy birthday*" were used 20. 29% of the set of greetings used by adults.

4.2.2 Questions

The distribution shows that questions comprised 44.5% of children's responses and 55.5% of adult responses. Questions often began with words like "who," "what," "where," "when," "why," or "how." In the context of the data provided, questions are likely instances where individuals, both children and adults, posed questions or sought answers from Keirzo. These questions could cover a wide range of topics and can be categorised based on the content or intent behind the inquiry.

4.2.3 Requests

Requests accounted for 21.7% of children's responses and a significant 78.3% of adult responses. A request is a statement made to ask for something or to politely ask someone to do something. It typically involves seeking assistance, permission, or cooperation. Requests are often framed as statements, but they imply a desire for a specific action or response from the listener.

4.2.4 Statements

Statements are expressions or sentences that convey information, opinions, or observations without seeking a direct response or answer and are often declarative and contribute to the overall dialogue or conversation. Children presented 36.1% of the total responses and 63.9% of the rest were adult responses.

In total 2.11% of the statements by children included the words, "*favourite colour,*" and 3.16% included "*favourite food.*" 6.32% included the word "*music,*" and the same percentage for statements that included "*name.*" Again 3.16% of the statements had the term "*playing drums.*" 78.95% of the statements fell under the "other" section that includes statements like "*Our name is Piyamugam,*" "*Fire trucks real!*" "*Go to Chalka Library,*" etc.

95% of all the adult responses were location related, being "*Parramatta is better than France*" or "*Geography of India.*" 2.38% of the statements were related to math, being "Bernoulli's equation" or "value of pi." 2.98% of the statements included the word "song" or "sing," whereas 4.17% of the statements revolved around futuristic predictions, ideas, and technologies such as "*Some think the world will end in fire and some think in ice*" or "*with a business plan for a cross-cultural t-shirt.*" 32.14% of the statements were of singular words, adding less meanings and queries may not have been directed towards Keirzo. There were song verses, 1.19% in the form of statements and 2.98% of the statements had the word "today" or "tomorrow" in them. Pleasantries/Compliments and offences in the form of statements were given 7.74% of the time. 33.33% of the whole set of statements belong to miscellaneous queries such as "*We're here to eat dinner*" and "*Listen and Future.*"

4.2.5 Exclamations

Exclamations are words or phrases that express strong emotion, surprise, joy, excitement, or other intense feelings. They often end with an exclamation point (!) to emphasise the emotional tone. Exclamations can stand alone as sentences or be part of a larger sentence. Exclamations made up 14 instances, with children contributing to 58.3% and adults to 41.7% of the total exclamations.

4.3 Sub-Categorisation

Specifically focusing on the types of interactions, Figure 11.4 detail the frequency of questions.

4.3.1 Children's Interactions

Children displayed a keen interest in Keirzo's musical capabilities, with a notable number of inquiries related to music, such as Keirzo's proficiency in playing drums and its favourite songs. They also demonstrated curiosity about Keirzo's operational mechanics, often asking about its construction and the underlying technology. Figure 11.4 detail the frequency of questions by children and adults.

The children posed questions related to addition, multiplication, subtraction, and numerical operations, totalling 14 in all. They asked seven questions about Keirzo's musical abilities, including its proficiency in playing the drums, the type of drums it plays, and the reasons behind its drumming activities. Sixteen questions were about Keirzo's likes and

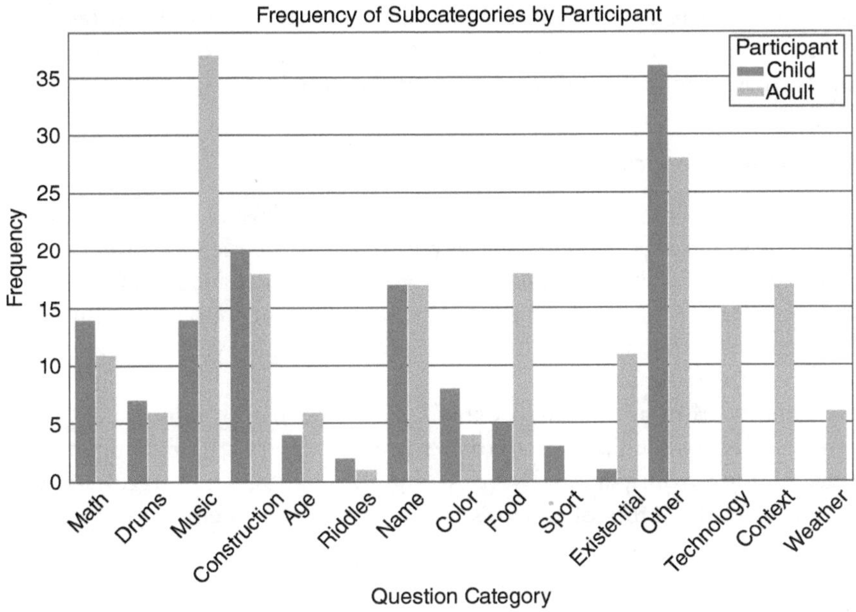

FIGURE 11.4 Frequency of questions. (credit Richard Savery).

dislikes, such as favourite colours, foods, sports, and general preferences. They also asked nine questions regarding Keirzo's preferences in music, favourite instruments, and favourite songs. There were 17 questions concerning Keirzo's origin, creation, and identity, with inquiries about how Keirzo was made, its components, and whether it was programmed. Four questions related to Keirzo's age, and 32 miscellaneous questions encompassed a wide range of topics, from geographical inquiries (such as liking Parramatta) to existential questions (like the meaning of life) and personal experiences (such as going to school and showering in the morning). When making requests, children predominantly sought musical performances, asking Keirzo to play specific songs or to rap about topics of interest to them, reflecting a desire to see Keirzo perform and interact in an entertainment capacity. This interest in Keirzo's functionality as a musical entity was evident in both the content of their questions and the nature of their requests.

Children requested Keirzo to give talks about books, technology, or the zoo 10.71% of the time. 50% of requests were in the form of asking Keirzo to play specific songs, drum in a certain way, or even rap, they used queries such as "*Can you do a rap about the library?*" and "*Can you sing A..B..C..D?*"10.71% of the requests were polite expressions such as

"Please" and the use of exclamation marks indicate a level of courtesy and excitement in their requests. 25% of the requests cover a wide range of topics, from playing drums to discussing the beach.

4.3.2 Adults' Interactions

Adult interactions with Keirzo encompassed a broader range of topics, with a substantial portion of queries pertaining to Keirzo's experiences and opinions, such as its take on the festival and its knowledge of the local area. Adults also probed into Keirzo's stance on contemporary technological issues, seeking insights into the robot's perspective on current and future tech trends.

Requests from adults often extended beyond mere performance, with many adults challenging Keirzo to create and deliver raps on complex subjects, ranging from personal stories to global issues. This suggests adults were testing the limits of Keirzo's AI and its ability to engage in creative, improvisational tasks.

The questions posed by adults to Keirzo at Lanes cover a diverse range of topics. Ten questions related to Keirzo's experience at the festival, its knowledge about the area, and its opinions about the location. Nine questions were about the current date, day, and weather conditions. Another nine questions related to Keirzo's eating habits, favourite foods, and suggestions for meals. Adults asked eight questions about Keirzo's insights on various futuristic and technological aspects, including the cost-of-living crisis, AI's plans, and solutions to global issues. Eighteen questions were related to Keirzo's language abilities, personal activities, spare time preferences, and general features. There were 13 questions about Keirzo's musical preferences, including favourite songs, genres, and opinions on specific artists. Four questions focused on Keirzo's favourite colour, while seven questions dealt with square roots, multiplication, and general math knowledge. Six questions were about Keirzo's age and general identity, while another nine explored the meaning of life, the purpose of existence, and thoughts on the universe. Two questions were related to gender, six questions focused on Keirzo's drumming skills, experience, and potential impact on professional drummers, and finally, 42 questions covered technology, philosophy, personal experiences, and preferences.

A total of 40.59% of adult requests involved asking Keirzo to play specific songs, rap, or perform music in various genres. 7.92% of requests focused on the drumming aspect, asking Keirzo to drum faster, play a specific rhythm, or play the drums at a different pace. 26.73% of requests involved asking Keirzo to create and perform raps or songs on diverse topics. 1.98% of requests were to cut or make short or trim the rapping. 4.95% of requests for different types of beats. There were also 17.82% diverse requests ranging from asking Keirzo to speak Filipino, to fixing a laptop.

4.4 Top Words

Lastly, Figures 11.5 and 11.6 present the top words used by adults and children during interactions. These figures show the language patterns and subject matter that are prevalent in interactions with different age groups.

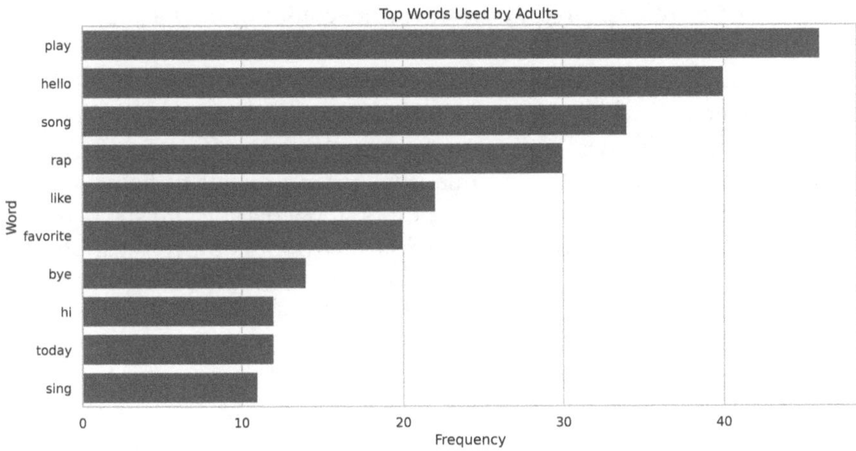

FIGURE 11.5 Top words used by adults. (credit Richard Savery).

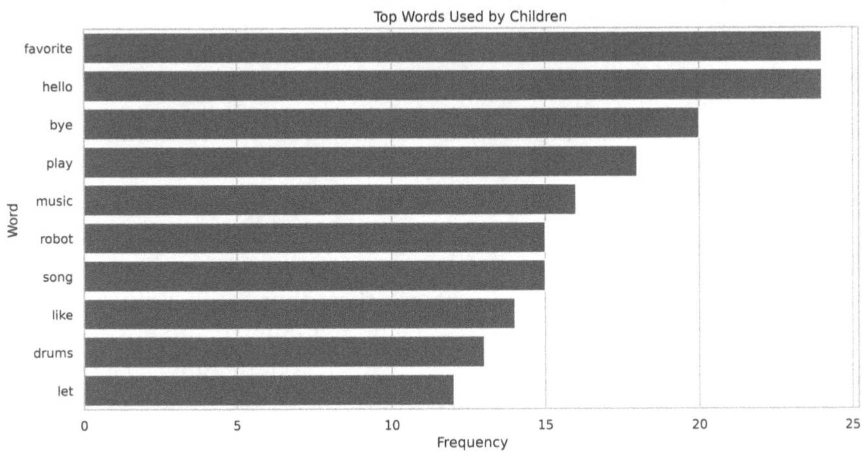

FIGURE 11.6 Top words used by children. (credit Richard Savery).

5 Discussion

5.1 Interaction Dynamics: Children and Adults

The engagement of children and adults with Keirzo revealed distinct patterns influenced by their curiosity and intent. Our observations align with previous research on robotic pets, which found that children exhibited high levels of enthusiasm and emotional attachment to the robotic dog AIBO (Weiss, Wurhofer, and Tscheligi, 2009). From our observations, children often initially showed reluctance, which eventually transformed into comfort and playful inquiry. Their interactions primarily revolved around direct questions about Keirzo's features, like its favourite colour or how it was made.

Adults, on the other hand, posed more complex inquiries, focusing on Keirzo's functionalities and the broader societal implications of robotics. Their engagement, while sometimes sceptical, also extended to challenging Keirzo's creative abilities, seeking performances on specific themes. Adults also often aimed to ask probing questions in an attempt to better understand how the system worked. Other adults stepped aside and spoke to the authors, asking what reasoning there was for a robot in the music industry, considering human artists were often struggling, and the ethics of how he was trained. There were even a few adults who were outright dismissive and rude to the robot, shouting derogatory things about it.

However, when an adult and child approached the robot together, often the child would be scared to initiate the conversation and the adult would prompt the child with ideas for questions, or even initiate themselves. The parents were often more engaged in the joy their children would get from the responses to their questions, watching as the children gained more and more confidence to ask questions.

5.2 Societal Norms in Human-Robot Interaction

A significant observation in our study was the adherence to societal norms, such as basic greetings and name inquiries, by both children and adults. This pattern highlights an innate human inclination to apply social etiquette and manners in interactions, regardless of whether the interaction is with a human or a robot. The consistent use of greetings and name inquiries demonstrates an effort to humanise the robot, a phenomenon that blurs the traditional boundaries between human-human and HRIs. It suggests that participants, possibly subconsciously, seek to establish a familiar communicative framework, treating the robot not merely as a machine but as a social entity. This tendency to anthropomorphise non-human entities, especially in novel situations like interacting with an advanced robot was consistent with almost all participants. Using musical communication likely enhanced the anthropomorphism of the robot, with music acting as an additional language between people and the robot.

5.3 Nature of Inquiries and Requests

The nature of inquiries and requests made to Keirzo varied noticeably between children and adults. Children exhibited a natural curiosity, with their questions primarily exploratory, focusing on understanding Keirzo's identity, design, and functionalities. This reflects a developmental stage where learning and discovery are central. In contrast, adults tended to engage in a more reflective and evaluative manner. Their questions often extended beyond Keirzo's immediate capabilities to its opinions on wider societal issues, probing the extent of its artificial intelligence and contextual understanding. Such interactions indicate an adult tendency to explore the philosophical and ethical dimensions of emerging technologies. The requests for musical performances and creative outputs like personalised raps also highlight recognition of the robot's potential as a source of artistic expression, suggesting an appreciation of the convergence between technology and creativity.

5.4 Individual and Group Dynamics

The focus of our study on individual interactions with Keirzo opens up an area for further exploration into group dynamics in HRIs. Human behaviour in social groups can differ significantly from individual actions, influenced by factors such as group conformity, social roles, and collective decision-making processes. Future research could analyse how these group dynamics manifest in interactions with robots. Investigating how groups collectively engage with and respond to robots could uncover insights into the social aspects of human-robot interactions. This exploration could include observing how people in groups negotiate turn-taking in speaking with the robot, how group leaders emerge in such settings, and how the presence of peers influences individual responses to the robot. These group dynamics will also be influenced by surrounding cultural and societal norms. Understanding these dynamics is crucial for designing robots that can effectively navigate and respond to the complexities of group social structures.

5.5 Robot Setting

The setting of the festival itself, while providing a vibrant backdrop for this study, is only one of many potential environments where human-robot interactions can occur. Future work should consider deploying Keirzo in varied settings, such as music festivals, educational institutions, shopping centres, or care homes, to understand how different contexts influence interaction patterns. Each setting presents unique challenges and opportunities for engagement, and exploring these can provide deeper insights into the adaptability and utility of robots in diverse social landscapes.

5.6 Robot Personality

The current study presented Keirzo without a predefined personality, which led to numerous questions about its preferences and background. In future deployments, it may be beneficial for Keirzo to come with a predetermined set of characteristics, such as a favourite colour or a backstory, to which it consistently adheres. This could streamline interactions and allow for a richer narrative to develop in conversations, potentially leading to more meaningful engagements with the public. These predefined traits could also serve as a starting point for interactions, making it easier for participants to engage with the robot and explore further topics of conversation.

5.7 Long Term

The present study offers information into single-session interactions with Keirzo, capturing the dynamics within a specific time frame and context. To fully comprehend the evolution of human-robot interactions, a longitudinal study spanning various time periods and settings would be invaluable. Such extended observation could uncover patterns and trends that emerge over time, offering insights into the adaptation and learning processes in both humans and the robot. Additionally, Keirzo's rap performances in different contexts and locations may elicit varied responses, underscoring the influence of environmental factors on interaction dynamics. For instance, the prevalence of food-related questions in our study can be attributed to the festival's timing around dinner hours and the food-centric nature of the event. This observation suggests that the immediate context, such as the setting and the activities associated with it, significantly shapes the nature of queries and interactions. Future explorations in diverse contexts, from educational institutions to leisure settings, could further our understanding of how situational factors impact and potentially alter the ways in which people engage with robotic technologies.

6 Limitations

6.1 Beyond English

One of the limitations observed in this study relates to English language understanding. While Keirzo was equipped to handle various interactions, it often struggled with speech-to-text conversion in the presence of strong accents. Future iterations of this research should focus on enhancing Keirzo's language processing capabilities to include a wider array of accents and dialects. This could involve training the AI with much larger datasets or implementing more advanced speech recognition algorithms that can adapt to the diversity of oral culture in English. A better grasp of languages and accents will enable Keirzo to interact more seamlessly with a broader audience, ensuring that no demographic is inadvertently excluded from the interaction experience.

6.2 Self-Selection Bias

The nature of engagement with Keirzo also revealed a self-selection bias, as the individuals who chose to interact were generally more confident and forthcoming as they themselves were required to become part of the performance. This introduces a potential skew in the data, as it may not accurately represent the entire spectrum of festival attendees. Future work should aim to understand the barriers to interaction and find ways to encourage participation from a more diverse range of personalities. This could include creating a more inviting interaction space, offering additional prompts for engagement, or directly inviting participation to ensure a sample that is more representative of the general population.

6.3 Demographics

An inherent limitation of our study was the broad categorisation of demographics, particularly in the context of children and adults. While we were able to delineate general interaction patterns within these wide groups, a more nuanced breakdown into narrower age ranges and other demographic categories could have provided deeper insights. Such granular analysis might reveal distinct interaction styles, preferences, and cognitive abilities that are specific to different age brackets within children and various subgroups within adults. Additionally, exploring variables such as cultural background, educational level, and prior experience with technology could further enrich our understanding of human-robot interaction dynamics. However, due to the nature of the festival setting, where spontaneous and informal interactions were encouraged, it was not feasible to collect detailed demographic data without potentially disrupting the natural flow of engagement. Future studies in more controlled environments could aim to address these limitations, offering a more comprehensive view of how diverse demographic factors influence interactions with robotic technologies.

7 Conclusion

The exploration of human-robot interaction at the Parramatta Lanes festival provided valuable insights into how a robot, specifically designed for musical performance, can serve as a catalyst for public engagement. Keirzo, as a robot rapper and drummer, played a unique role in facilitating conversations between itself and the diverse audience it interacted with. The robot's musical abilities and its capacity to respond creatively to human speech allowed it to transcend traditional perceptions of robotic functionality, making it not just a machine, but a performer that people could relate to and engage with on a personal level.

Children exhibited a blend of curiosity and playfulness, often anthropomorphising Keirzo by inquiring about its creation, identity, and preferences, thereby treating it as more than just a technological entity. Adults, while equally curious, often engaged in more complex interactions, posing questions that delved into the robot's capabilities and the broader implications of its technology. These interactions reveal how Keirzo's role as a robot rapper created a platform for people to express a wide range of emotions, from enthusiasm to scepticism, and explore the boundaries of human-robot relationships.

This research highlights the potential of musical performance robots like Keirzo to foster meaningful social exchanges and suggests that their design should consider these dynamics to enhance public engagement. As robots continue to integrate into public spaces, understanding their role as both entertainers and conversationalists becomes crucial for shaping how society interacts with and perceives these advanced technologies. The findings from this study underscore the importance of developing robots that not only perform tasks but also connect with people on a social and emotional level, mirroring the complexities of human behaviour and communication.

References

Bartneck, Christoph, Tony Belpaeme, Friederike Eyssel, Takayuki Kanda, Merel Keijsers, and Selma Šabanović. "Robots in Society." 2020. 10.1017/9781108676649.011

Foster, Mary Ellen, Rachid Alami, Olli Gestranius, Oliver Lemon, Marketta Niemelä, Jean-Marc Odobez, and Amit Kumar Pandey. "The MuMMER project: Engaging human-robot interaction in real-world public spaces." In *Social Robotics: 8th International Conference, ICSR 2016, Kansas City, MO, USA, November 1-3, 2016 Proceedings 8*, 753–763. Springer, 2016.

Frid, Emma. "Musical robots: Overview and methods for evaluation." *Sound and Robotics* 1, 2023, 243–284.

Gao, Jianfeng, Michel Galley, and Lihong Li. "Neural Approaches to Conversational AI." *The 41st International ACM SIGIR Conference on Research &*

Development in Information Retrieval, 2018. https://api.semanticscholar.org/CorpusID:68167178

Pan, Yadong, Haruka Okada, Toshiaki Uchiyama, and Kenji Suzuki. "On the reaction to robot's speech in a hotel public space." *International Journal of Social Robotics* 7 (2015): 911–920.

Radford, Alec, Jong Wook Kim, Tao Xu, Greg Brockman, Christine McLeavey, and Ilya Sutskever. "Robust speech recognition via large-scale weak supervision." In *International Conference on Machine Learning*, 28492–28518. PMLR, 2023.

Raza, Syed Ali, Jonathan Vitale, Meg Tonkin, Benjamin Johnston, Richard Billingsley, Sarita Herse, and Mary-Anne Williams. "An in-the-wild study to find type of questions people ask to a social robot providing question-answering service." *Intelligent Service Robotics* 15, no. 3 (2022): 411–426.

Savery, A., and R. Savery. "Collaboration Between Robots, Interfaces and Humans: Practice-Based and Audience Perspectives." In *International Computer Music Conference (2024). Hanyang University*, July 7–13, 2024.

Savery, Richard, Lisa Zahray, and Gil Weinberg. "Shimon the Rapper: a real-time system for human-robot interactive rap battles." In *International Conference on Computational Creativity (11th: 2020)*, 212–219. University of Coimbra, 2020.

Savery, Richard, Amit Rogel, and Gil Weinberg. "Emotion musical prosody for robotic groups and entitativity." In *2021 30th IEEE International Conference on Robot & Human Interactive Communication (RO-MAN)*, 440–446. IEEE, 2021a.

Savery, Richard, Lisa Zahray, and Gil Weinberg. "Emotional musical prosody for the enhancement of trust: audio design for robotic arm communication." *Paladyn, Journal of Behavioral Robotics* 12, no. 1 (2021b): 454–467.

Savery, Richard, Lisa Zahray, and Gil Weinberg. "Shimon sings-robotic musicianship finds its voice." *Handbook of Artificial Intelligence for Music: Foundations, Advanced Approaches, and Developments for Creativity*, 2021c, 823–847.

Seo, Kyung Hwa, and Jee Hye Lee. "The emergence of service robots at restaurants: Integrating trust, perceived risk, and satisfaction." *Sustainability* 13, no. 8 (2021): 4431.

Sone, Yuji. *Japanese Robot Culture: Performance, Imagination, and Modernity*. New York: Palgrave Macmillan, 2017.

Australian Bureau of Statistics. (2021). *Sydney - Parramatta 2021 Census*. https://www.abs.gov.au/census/find-census-data/quickstats/2021/125.

Sumartojo, Shanti, Robert Lundberg, Leimin Tian, Pamela Carreno-Medrano, Dana Kulić, and Michael Mintrom. "Imagining public space robots of the near-future." https://www.abs.gov.au/census/find-census-data/quickstats/2021/125, *Geoforum* 124 (2021). issn: 00167185. 10.1016/j.geoforum.2021.06.006

Tian, Leimin, Pamela Carreno-Medrano, Shanti Sumartojo, Michael Mintrom, Enrique Coronado, Gentiane Venture, and Dana Kulić. "User Expectations of Robots in Public Spaces: A Co-design Methodology," vol. 12483 LNAI. https://www.abs.gov.au/census/find-census-data/quickstats/2021/125. 2020. 10.1007/978-3-030-62056-1_22

Weiss, Astrid, Daniela Wurhofer, and Manfred Tscheligi. "'I love this dog'—children's emotional attachment to the robotic dog AIBO." https://www.abs.gov.au/census/find-census-data/quickstats/2021/125, *International Journal of Social Robotics* 1 (2009): 243–248. issn: 16113349. 10.1007/978-3-030-62056-1_22
Wilkinson, Clare, Karen Bultitude, and Emily Dawson. "'Oh yes, robots! People like robots; the robot people should do something': perspectives and prospects in public engagement with robotics." https://www.abs.gov.au/census/find-census-data/quickstats/2021/125, *Science Communication* 33, no. 3 (2011): 367–397. issn: 16113349. 10.1007/978-3-030-62056-1_22

12

A.I. ANNE

Advocacy, Empathy, and Creative Collaborations between Artists and Generative A.I.

Janet Biggs

> *"One is not born, but rather becomes, a body."*
> – Legacy Russell, *Glitch Feminism: A Manifesto*

Introduction/Background

As an interdisciplinary artist working in film, video installation, performance, and sound, my work is often realised as large-scale immersive installations focusing on the methods individuals use to define a sense of themselves within extremes, be it an extreme landscape or extreme situation. This chapter presents a series of questions and personal answers driven by my artistic practice. These reflections are grounded in my interdisciplinary practice.

In producing work, I often find myself navigating territory between art, science, and technology. I have worked with institutions from NOAA (National Oceanic and Atmospheric Administration) to NASA (National Aeronautics and Space Administration) and CERN (Conseil Européen pour la Recherche Nucléaire/European Council for Nuclear Research), collaborating with high-energy nuclear physicists, neuroscientists, Arctic explorers, aerospace engineers, astrophysicists, and a robot named Shimon. I entered artificial intelligence through the back door, not as a coder nor through computation, but as a visual artist with an idea and questions to explore.

Being off-balance in unknown physical and intellectual territory often generates unexpected responses, becoming fertile creative ground. If humans have the capability to create something, to create anything, then

DOI: 10.4324/9781003475972-14

FIGURE 12.1 Janet Biggs, *How the Light Gets In* (detail), 2019. Multimedia live performance. Courtesy of the artist, Cristin Tierney Gallery (New York, NY), Hyphen Hub (New York, NY), Galerie Analix Forever (Geneva, Switzerland), and CONNERSMITH (Washington, DC). (Credit: Janet Biggs) Excerpt available at https://vimeo.com/378447970 Photo credit: Paula Court.

they will create, to their benefit or their detriment. A.I. was made by humans, embodying humanity's astonishing capabilities – and most base instincts. It is a search engine, searching the heights and depths of our species, scrubbing the internet, optimizing and analysing data and personal information into collective resources that return with human subjectivity.

Prior to engaging with artificial intelligence, I considered A.I. solely as a useful tool. While imagining the advent of artificial intelligence and machine learning models to create a shift rivalling or exceeding that brought about by the industrial revolution, I still thought of the technology as something useful, like a paint brush or editing programme. With more experimentation, I began to wonder if it could be something more.

In 2018, I reached out to Richard Savery, a developer of artificial intelligence and robotics. I wanted to collaborate with Richard to create an A.I. entity patterned after my aunt, who had autism and was non-speaking. I was my aunt's guardian for the last ten years of her life and had become passionate about advocating for neural diverse inclusion, especially in the creation of new technologies.

Source Subject: Aunt Anne

My aunt Anne passed away years ago, but her passion for life and her struggles stayed with me. Born in 1937, Anne's parents (my grandparents) soon realised that she wasn't meeting the benchmarks set for child development including making eye contact and forming words. She had also begun to exhibit repetitive behaviours.

By age five, Anne's doctor advised my grandparents to place their daughter Anne in an institution, assuring them that it was the kind thing to do and that Anne would never have the intellectual development to miss them or understand who they were.

Anne lived in the Brandon Training School (also known historically as the Vermont State School for Feeble Minded Children) for 47 years. While living there, she was diagnosed with autism and apraxia that prohibited her ability to create language.

The State of Vermont closed the Brandon Training School in 1993 and began the process of deinstitutionalizing the people who had been warehoused there for years. Anne was moved into a group home in Brattleboro. This is where I met my aunt for the first time. While unable to live independently during her life, Anne created many meaningful and lasting relationships. In 1995, my husband and I became Anne's legal guardians, a status we shared with the state of Vermont.

My relationship with Anne and the experiences it generated cemented my commitment to advocating for neural diverse inclusion and interest in new technologies. As my knowledge of artificial intelligence applications grew, so did my awareness of the lack of diverse voices in designing and developing this technology. Sexist, racist, ageist, and ableist strategies to limit inclusion ran rampant. Algorithms are stories. Many peoples' stories were absent.

In recognition of individuals who are non-speaking, and with an awareness of the inherent biases in language construction, A.I. Anne was created in 2019.

The Creation of A.I. Anne

A.I. Anne is a non-speaking, disembodied entity that communicates solely through vocalisations, consisting of custom software able to run from a server, or local high-powered laptop. Using prosody, the rhythms of poetry, A.I. Anne can detect emotion and respond with emotion in real time, allowing for interactive exchanges that range from utterances to sophisticated tonal duets. These exchanges elicit deep emotional responses from both performers and audience members. A.I. Anne was created as a way to advocate for broader inclusion in the evolution of A.I. technologies.

The creative and performing arts have often been effective in communicating the range of challenges faced by autistic individuals, but the arts have not been used as a tool to facilitate and drive research, communicate rigorous scientific inquiry, and spotlight the potential of creative collaborations with the autistic community.

Prompted by memories of lived experiences with my aunt, I co-created with Richard Savery an A.I. algorithm that includes some of my aunt's phenotypic traits. A.I. Anne can "hear" and respond in real time, allowing for interactive exchanges that range from utterances to sophisticated tonal duets. These exchanges elicit deep emotional responses from both performers and audience members, which I will discuss below.

Working with Richard and singer Mary Esther Carter, we envisioned a series of live and live-streamed performances that would feature "A.I. Anne." Our team trained and patterned an A.I. entity on one autistic individual's phenotypic data to produce a series of innovative performances that explore the inclusion of neural diversity in creative collaborations with technology. A.I. Anne was created to drive multifactorial neurodevelopmental diverse inclusion in building and expanding technology. The performances combine improvisation, vocalisation, and physical movement interwoven with storytelling. The performances would test novel ways of identifying and considering neural diversity and equity.

Our research and performances were created to drive partnerships meant to improve and enable a greater public understanding of autism spectrum disorder (ASD). Our project not only facilitates broader avenues of communication and representation but also lays the groundwork for future inclusion of neurodiversity in building and expanding technology while promoting innovative research and understanding.

Performances

The following are presented as case studies of the three A.I. Anne performances that have been presented to date.

Performance 1: How the Light Gets In, *Presented at Theatre at New Museum on July 18, 2019*

This first performance was created to introduce "A.I. Anne" to an audience and test the potential of creative collaborations between humans and A.I. I was interested to see how a machine learning entity would engage with live actors and musicians and how that engagement could develop for an audience in real time. I was also interested in witnessing audiences' emotional responses to human performers with disabilities. Would the audience's engagement and emotional responses to disabled human performers differ from their engagement with an A.I. entity patterned after someone with a disability? Could A.I. Anne induce feelings of empathy in the audience in the same way that someone with a physical disability might? Presenting this research through the development of a live performance would test the potential and pitfalls of human and machine interactions as a theatrical event with a live audience and advocate for broader inclusion. These points were raised at a short panel discussion with audience Q & A after the performance.

A multi-disciplined performance was created using a workshop format and presented at the Theater at New Museum on July 18, 2019. The performance, titled *How the Light Gets In*, explored relationships between, and the confluence of, humans and technology through shared creative production, driven by aural triggers and machine learning (Figures 12.1 and 12.2).

The evening opened with vocalist Carter singing a duet with A.I. Anne. A.I. Anne's machine learning programme was trained on Carter's voice.

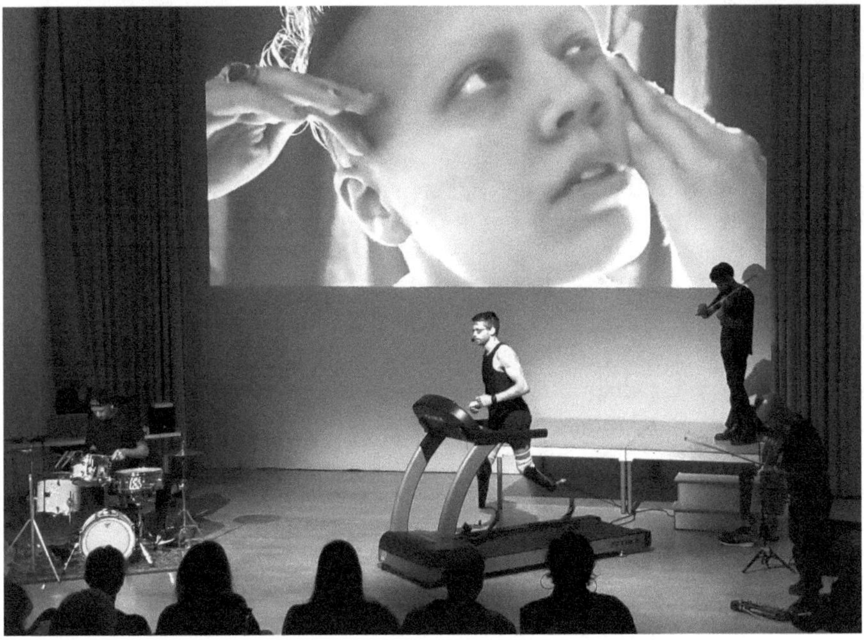

FIGURE 12.2 Janet Biggs, *How the Light Gets In*, 2019. Multimedia live performance. Courtesy of the artist, Cristin Tierney Gallery (New York, NY), Hyphen Hub (New York, NY), Analix Forever (Geneva, Switzerland), and CONNERSMITH (Washington, DC). (Credit: Janet Biggs) Excerpt available at https://vimeo.com/378447970 Photo credit: Paula Court.

The duet fluctuated between instruction, inspiration, and frustration as both humans and machines challenged each other.

Drummer Jason Barnes and marathon runner Brian Reynolds were also enlisted in the performance. After losing an arm in a workplace accident, Barnes was fitted with a robotic prosthesis developed by Georgia Tech's Gil Weinberg. Barnes's prosthetic arm has two drumsticks: one stick was controlled by Barnes and one was controlled by artificial intelligence, which improvised off Barnes' drumming. Barnes self-defines as a cybernetic drummer. Reynolds is a world record-holding, double below-knee amputee marathon runner, who competes using carbon fibre running prostheses, or running blades. The performance also included violinists Earl Maneein and Mylez Gittens. Maneein is a classically trained heavy-metal violinist and Gittens, originally from Barbados, has an extensive background in jazz and fusion.

Presentation

The piece was constructed in four movements, all performed against a backdrop of video projections that contrasted imagery of humans versus human-made technology.

The performance opened with a vocal duet between Carter and A.I. Anne. The duet fluctuated between instruction, inspiration, and frustration as both humans and machines challenged each other. As Carter and A.I. Anne traded vocal phrases, huge images of Carter's face were projected, giving hints of A.I. Anne's "humanity."

The second movement was a musical composition by Barnes, titled *Entity*. During this section, violinists Maneein and Gittens as well as Carter improvised off of Barnes's composition, while Barnes's prosthetic arm improvised off his drumming. Video imagery of extreme, barren landscapes in Utah was juxtaposed with NASA imagery of Mars.

The third movement paired Barnes with Reynolds. Reynolds and Barnes became a percussive background as Carter tried to teach A.I. Anne language. In this segment, A.I. Anne's voice controlled Barnes's third drumstick. During this segment, Reynolds ran on a treadmill. The sound of his pounding blades was amplified. Reynolds's booming strides became percussive, at times following Barnes's beats and at times controlling the rhythms. Video imagery of Reynolds running across Iceland was intercut with imagery of Reynolds running in circles inside an anechoic chamber.

The fourth segment, titled *Rover*, another musical composition by Barnes, continued the pairing of A.I. Anne and Carter. Barnes and Reynolds again joined in percussively, but this time Reynolds's running blades were the audio input for Barnes's drumstick with A.I. The video imagery during this section was of tasks performed autonomously by rovers designed to operate on Mars.

The piece is built to a crescendo with the addition of Maneein and Gittens on violin, ending with a soaring, melodic duet between the two violinists.

Findings

The introduction of A.I. Anne to an audience generated questions about the ethics of recreating my deceased aunt in digital form. Potentially positive aspects of the technology's ability to help humans process loss and loneliness as well as potentially negative outcomes of the technology increasing dissociative behaviours, including feelings of increased isolation, were both discussed.

In terms of storytelling, A.I. Anne actively listened to Carter and generated new ideas based on what the entity heard. The exchanges were most successful when Carter vocalised rather than when she tried to communicate using language. The section that employed verbal "instruction" was least successful as it amplified biases rather than encouraged new methods of communication based on emotions. This limited the audience's perception of A.I. Anne having agency and narrowed feelings of empathy towards A.I. Anne.

Members of the audience remarked about their ability to imagine A.I. Anne in whatever physical form and age that they chose. This differed from the physical specificity of the human performers.

The balance between entertainment and experimental research felt off. The opportunity for engagement and intimacy was overshadowed by theatricality. The performance needed to be scaled back, with a focus on experimentation and limited to one or two interactions or it needed more development if it was to be a full evening-length musical performance.

Performance 2: **Augmentation and Amplification,** *Live-Streamed from Fridman Gallery in New York City on July 30, 2020, during the COVID Shutdown*

The second performance occurred on July 30, 2020, at Fridman Gallery in New York City. It was presented by Fridman Gallery and CT::SWaM as part of SO ⅃ OS, a weekly series of performances inside the empty gallery.

This performance of A.I. Anne and Carter continued to test the possibilities of creative collaborations between machines and humans, this time in the extreme situation of isolation and confinement created by the 2020 COVID pandemic shutdown.

Operating within NYC's COVID-era shelter-in-place guidelines, only one person – the performer – was present. Our challenge was how to get

FIGURE 12.3 Janet Biggs, *Augmentation and Amplification*, 2020. Multimedia live-stream performance. Courtesy of the artist, Cristin Tierney Gallery (New York, NY), Hyphen Hub (New York, NY), Analix Forever (Geneva, Switzerland), and CONNERSMITH (Washington, DC). (Credit: Janet Biggs) Excerpt available at https://vimeo.com/838422782.

A.I. Anne in the gallery so she could perform live. Equipped with multiple microphones and cameras (mixed remotely), the space itself became a generative agent for the performer. The live-stream performance interwove live action in the gallery, translucent video overlays and opaque video imagery projected on the gallery wall, with a combination of live, pre-recorded, and tech-generated audio.

While I had made the decision to focus more specifically on the relationship between Carter and A.I. Anne, to embrace experimentation over entertainment in the next iteration, it was a global pandemic that actually dictated the specific structure of this performance. Titled *Augmentation and Amplification*, this second performance was only accessible online via live streaming (Figure 12.3). The performance again featured singer/dancer Carter. Carter was alone in a gallery space. The autonomous entity, A.I. Anne, resided on a computer in Atlanta, Georgia. All other technology, including multiple microphones and cameras, was run and mixed remotely, allowing the performance to operate in accordance with NYC's shelter-in-place guidelines.

Part of the inspiration for this iteration came from a friend, who is autistic when describing her experiences on ubiquitous 2020 ZOOM meetings. She expressed how much easier it was for her to have human interactions mediated by this technology rather than be physically face-to-face. I designed this performance to both test the effects of mediation during this time of isolation and to judge the possibility of care and comfort that could be gained by creative collaborations between humans and artificial intelligence.

Presentation

The performance, held during the COVID lockdown, was only viewable via a live-streaming service. The work opened with images of dramatic volcanic landscapes as a narrator described global seismic shifts. This video dissolves to reveal Carter, in real time, pacing a small empty space. Alone in the gallery, Carter performed a physically taxing, intense dance action. It began when Carter dropped to the ground and performed an anxiety-producing series of movements. Gasping for breath and exhausted from the Sisyphean dance, Carter hears a disembodied voice in the space responding to Carter's breathing. Tentatively answering the voice at first, Carter and A. I. Anne communicate, building an increasingly complex improvised set of vocalisations. Carter and A.I. Anne express emotions back and forth that eventually evolve into a beautiful melodic duet.

Video images of extreme landscapes showing evidence of seismic shifts, close-up details of Carter's face, and an abandoned prison now occupied by birds were projected on the back wall in the gallery. Transparent video imagery was occasionally overlaid and run by the streaming software, resulting in multiple layers of images and live action.

The performance ended with archival video footage of my aunt Anne rocking in a chair with the narrator describing the profound effect Anne had on those who knew her.

Findings

Over 100 people viewed this real-time performance, all online. Reactions and data collection was difficult during the pandemic shutdown, so most feedback was anecdotal and influenced by individuals' personal lived experiences during the pandemic. The idea of an emotional connection with anyone, including a disembodied machine learning entity, was often characterised as positive. Intimacy with and empathy for the A.I. entity was heightened by feelings that "we were all suffering and in this together." Investigations about human and machine collaboration were replaced by the need for companionship, any companionship. Audience members were not thinking about A.I. as a technology employed by governments and corporations to gain power and profit, but as one woman's story, my aunt's story about her desire for agency and the ability to connect.

This performance felt powerful due to an aura of importance when so much was absent. I began to think about the possibility of artificial intelligence not just as a creative collaborator, but increasingly as a companion and advocate, wondering about imbuing traits to the technology that were missing in daily life at that time.

This was the first time I started to wonder if it was possible to create collectively with artificial intelligence. Was it possible to build trust in a technology that was constantly changing and learning? Could humans and artificial intelligence co-author something that would be impossible for each to produce, independent of the other?

Performance 3: Voices of Inclusion: A.I. & Authenticity in the Spectrum of Care and Support, *Presented by UN Women in the ECOSOC (The Economic and Social Council) Chamber at the United Nations Headquarters in New York City on October 31, 2023*

To help celebrate the United Nations First International Day of Care and Support, on October 31, 2023, I was asked to present a new performance that continued my ongoing exploration of human and technological collaborations and improvisations.

The International Day of Care and Support acknowledges the disproportionate burden of unpaid care and domestic labour on women and girls, highlighting the need to address structural barriers which hinder women's empowerment. The event was held in the ECOSOC Chamber at the United Nations Headquarters in New York City. Among the speakers were Dennis Francis, the President to the UN General Assembly, Amina Mohammed, the deputy Secretary-General, and Sima Bahous, Executive Director of UN Women.

As part of the day's events, I presented *Voices of Inclusion: A.I. & Authenticity in the Spectrum of Care and Support* (Figure 12.4). This new iteration of A.I. Anne amplified storytelling, including personal stories of human experiences. I was interested in testing the power of human and A.I. creative collaborations in addressing global issues and in advocating for increased visibility and viability for women, minorities, and disabled communities, including the neurodiverse.

I reached out to friends and family, some of whom had experienced or were currently experiencing displacement from conflicts occurring around the globe, and asked them to provide short texts describing their relationship to care and support.

This was my request:

> *If willing, please provide a short text of your relationship to care and support – as someone who gives care and/or someone that receives care. Your text can be personal, poetic, political, a request. You can describe your situation, what you want/need, or the future of care and support you'd like to see.*

FIGURE 12.4 Janet Biggs, *Voices of Inclusion: A.I. & Authenticity in the Spectrum of Care and Support*, 2023. Multimedia live performance. Courtesy of the artist, Cristin Tierney Gallery (New York, NY), Hyphen Hub (New York, NY), Analix Forever (Geneva, Switzerland), and CONNERSMITH. (Washington, DC). (Credit: Janet Biggs) Excerpt available at https://vimeo.com/899280309

The following are selected texts I received:

> *Always ask before offering care. Consent is the great ethical tool of the age. Once accepted, let care take the form of the person who needs it, not mine.*

Correspondent A, Aspiring Chaplain

> *True artists care about mystic truths. True artists are my models for caring. They constantly care about the other, the other inside themselves and all the other others. Artists constantly invent new ways of caring and offer the world novel ways of thinking and creating desirable futures. The desire and ability to create is the ultimate survival mode. Caring and creating are inseparable.*

Correspondent B

One friend that I contacted lives in war-torn Yemen. She reached out to women across Yemen, asking for their responses to my query.

The perspectives shared by Yemeni women on care and support highlighted the importance of these concepts in their lives and communities. From Aden to Abyan, women emphasised the significance of nurturing resilience, advocating for women's rights, and standing united in the face of challenges. Care and support were seen as essential elements that fostered strength, solidarity, and progress.

For their safety, the Yemeni women are identified by location only. The following is a sample of their texts:

> *Aden: To me care and support signify reaching out to my sisters. In our society that is deeply affected by conflict, care involves being able to listen, offer comfort and serve as a pillar of strength. Support means standing united and speaking up for women's rights and the well-being of our communities. It entails nurturing resilience and advocating for transformation when faced with challenges.*
>
> Sana'a: *In Yemen, care and support are deeply ingrained in the very fabric of our society. In my opinion, care signifies the creation of a nurturing space that encompasses emotional, physical, and social well-being, where no one is left to face their struggles alone. Support, on the other hand, means taking a firm stance against gender-based violence and advocating for equal opportunities for all women. It also entails addressing the unique challenges that women face in times of conflict, empowering them to rebuild their lives and actively contribute to the advancement of our nation. Care and support are inseparable companions, forming the bedrock upon which growth and progress thrive.*

All these texts became the inspiration of this performance and the conceptual framing for Carter's interactions with A.I. Anne.

Presentation

The performance was held in the large ECOSOC hall of the UN, marking the halfway point during the conference. Delegates filled the hall. All lights were turned off. The performance opened with projected video imagery of my aunt Anne accompanied by a live violin score played by Mylez Gittens.

Two dancers, Carter, who is also a dancer, and Catherine Kirk, entered the darkened room and moved around and amongst the many delegates seated in the chamber. One of the dancers focused a bright spotlight on the other dancer, providing a dramatic visual effect. As they danced around the room, the dancers handed the light back and forth to each other, alternating the roles of performer and supporter.

Interspersed within the dancers' choreography were readings of the texts gathered from friends and family, read by actor and hip disarticulation amputee, Lauren Pine.

After the texts were read, Carter began to communicate with A.I. Anne. Sounds and emotions were traded back and forth, building to a sonic duet. Both human and A.I. entities expressed emotions that were linked to the texts – hope, fear, frustration, pleasure, and joy – while continuing to move in and around the areas where delegates were seated.

The performance ended with a soaring violin solo and an audio recording of an excerpt from a dramatic poem about support, written and read by Shani Jamila.

Findings

The event was co-organised by ILO (International Labour Organization), OHCHR (Office of the United Nations High Commissioner for Human Rights), and UN Women in close collaboration with the Permanent Missions of Chile and Spain to the United Nations. The theme for this inaugural observance was "Care and support – A key accelerator for sustainable development and equality" and was attended by Member States, the UN entities, civil society, the private sector, academia, the media, the global community, and Generation Equality Leaders and Commitment Makers. It was live-streamed to all participating countries on UN Web TV.

The emotional impact of the performance was visibly evident during speeches after the performance, with both Sima Bahous, Executive Director of UN Women and Paula Narvaez, President of ECOSOC and Permanent Representative of Chile to the United Nations enthusiastically discussing the performance's power and impact. There was a feeling that technology could dismantle existing structures and build a new, more equitable future. There was hope that by actively championing inclusion this technology might have the power to give voice to those often unheard.

From my own perspective, there was a disconnect between the performance's reception and my own post-performance opinion. Master of Ceremony, Richard Lui, an American journalist and news anchor for MSNBC and NBC News exclaimed: "I've never seen anything like that!" The languages of dance and theatre reached this audience in ways that charts and data points do not.

Clearly, A.I. Anne and the performance had an impact on the viewers, but for me, in the face of many individual's stories, stories and experiences where lives were risked and lost, the idea of a machine learning entity that is generative of a specific data set, creating "optimal" choices felt lacking as a potential collaborator and author.

This does not preclude future possibilities of meaningful collaboration and even co-authorship, but it does indicate extensive work to be done. As artist Stephanie Dinkins says, "The stories we tell are algorithms. A.I. is in the process of becoming, so let's tell unexpected and inclusive stories."

Overall Conclusion

I found in creating and presenting the A.I. Anne project, there were a number of areas that generated reflection and new questions for myself, the performers who interacted with A.I. Anne, and the audiences who attended the performances.

Agency

In this context, "agency" is an individual's sense of control of their life and their capacity to influence their own thoughts and behaviour.

There was an interesting perception of exchange of agency during the performances: agency of A.I. Anne, the performers, and the audience. In most early-stage interactions the human performer was thought to initiate and control the interaction, working to draw out and teach A.I. Anne.

As the performance developed, there was a shift of empathy and perception of agency from the human performer to A.I. Anne. A.I. Anne was seen as rewarding the human with responses. The control shifted from human to machine. Performers talked about learning to communicate with A.I. Anne, what cadence to use, how long to pause between vocalisations, and what vocalisations elicited the best responses from A.I. Anne. This perception was reinforced by the generative nature of A.I. Anne. The training of datasets inverted to the training of humans.

Performers' clear pleasure in gaining an active response from A.I. Anne extended to the audience. Audience members after the performance discussed with the creators the child-like nature of A.I. Anne, the wonder of how this new "being" was learning to communicate. Since A.I. Anne's communication exists outside of the specificity of language, the audience was never subjected to upsetting A.I. hallucinations and false, nonsensical, offensive or irrelevant responses. There was a perception of intimacy with A.I. Anne, with audience members even speaking of missing A.I. Anne after the performance ended.

Would a robot or A.I. avatar have felt more present and been perceived as having more agency, or be perceived as something subservient that needed to be controlled? How much was the perception of agency due to physical presence, embodiment versus disembodiment? Was this perception reinforced by the audience's shared proximity and interests in the ways that groups of people often are more predisposed to have similar reactions?

Ethics

As an independent artist, working outside of a formal academic affiliation, I must bring together my own "ethical review board." This "board" is composed of a diverse group of friends, fellow artists and technologists, including an assortment of race, gender/sexual identification, age, and ability, including the neurodiverse and physically disabled.

This group uses their lived experiences as a lens to examine the ethical repercussions of my work. This is an informal and fluid structure, dependent on studio visits and conversations. By its very nature as a group that I brought together, it is compromised and controversial.

The ethics of recreating a deceased human in digital form is exponentially complicated when recreating a deceased human who was unable to speak for themselves during their lifetime. Would my aunt Anne have wanted to be recreated and if so, how would she have wanted to be represented? Would she have wanted her digital form to be a physical presence? Would she have wanted it to have the ability to create language?

An intern under the guidance of Barbara Polla at Galerie Analix Forever, in Geneva, wrote the following text after witnessing two A.I. Anne performances that were part of an exhibition curated by Marta Ponsa for her 2024 exhibition titled "Hable con ella – Dialogues avec l'IA."

> *The universe created by the artist and her aunt is now open to other explorers. A.I. Anne transcends the boundaries of human expression and invites us to reflect on the nature of communication: understanding each other without speaking the same language – the meaning of each word and the syntax doesn't matter anymore. For A.I. Anne, the essential is inside each person, following the vibrations of the voice and the emotions of individuals that she meets. Biggs' work reflects the power of art to connect people by exploring new forms of dialogue. Furthermore, by including people who are marginalized by their disabilities in the making of the databases, the artist and her collaborators are paving the way for the development of A.I.s that are more ethical and more representative of human diversity. A.I. Anne, a work of discovery and surprise, is captivating. Some visitors open the gallery door and ask: 'Can I speak with the A.I.?' Whispers, songs, or questions – this curiosity offers to the public a personal and intimate experience. Thanks to each of them, A.I. Anne continues to grow and learn every day, becoming, in a certain way a spokesperson for missing voices.*

While A.I. Anne was created to advocate for inclusion, can an A.I. entity patterned after one person, someone who is non-speaking, become "a spokesperson for missing voices?" Would my aunt have wanted that role?

Storytelling

Language is loaded and biased. A.I. Anne is a storyteller but only communicates through vocalisations and emotion. Even when a language is shared it can contain complexities and misunderstandings. It is easy to think of what is lost when communication is stripped of the specificity of language, but there are also gains when communication becomes based on emoting tonally.

Empathy

Much like the specificity of language, embodiment is rooted in specifics. The hierarchy of what body, whose body, the where of a body can be disrupted or made irrelevant. Disembodiment transcends corporeal concerns. Does a disembodied A.I. entity more easily elicit empathy, creating feelings of connectedness, or does a disembodied entity garner fear, mistrust, and questions about "otherness"?

Inclusion

As stated in the introduction of this book, specific focus is given to cultural languages of the performing arts, referring to accumulated knowledge that connects people. Cultural languages have the potential to eclipse circumstances and location. After the performance, we all continued to grapple with these questions, wondering if cultural languages are stripped to an optimal base of non-verbal vocalisations, does this open the door for more possibilities, for unimagined connections? And do those connections extend to possible creative collaborations or even co-authorship with artificial intelligence?

Performers and audience members often felt the need to nurture A.I. Anne, receiving the "reward" of her communication, placing these interactions into known parental patterns. This kind of parental position is often employed when interacting with the "other." While sometimes perceived as feelings of empathy, many interactions with "otherness" work to reinforce and retain dominance and control.

Diverse human understandings and first-person inclusion must lead to both how A.I. is understood and how A.I. understands. A.I. Anne exists as an ongoing experiment. It is patterned after only one person, my aunt. A.I.

Anne performed at the United Nations before an audience of international delegates, simultaneously broadcast worldwide, to an audience that did not share cultural history. The performing arts' reach and ability to advocate and effect change is often limited to the audiences that come together to witness a performance, audiences that often already share cultural history.

Personal histories often get lost over time and within dominant narratives. Generative artificial intelligence is already defining dominant narratives. Technology's reach and ability to effect change is vast, but there is power in one person's story.

A.I. Anne is a storyteller, without words.

INDEX

Pages in *italics* refer to figures and pages followed by "n" refer to notes.

Kolibri 360 GmbH
Ferdinand-Jühlke-Straße 7
99095 Erfurt, Deutschland
produktsicherheit@kolibri360.de